Mechanism Design

Mechanism design is an analytical about exactly what a given institutio to make decisions is dispersed and privately held. This book provides an account of the underlying mathematics of mechanism design based on linear programming. Three advantages characterize the approach. The first is simplicity: arguments based on linear programming are both elementary and transparent. The second is unity: the machinery of linear programming provides a way to unify results from disparate areas of mechanism design. The third is reach: the technique offers the ability to solve problems that appear to be beyond the reach of traditional methods. No claim is made that the approach advocated should supplant traditional mathematical machinery. Rather, the approach represents an addition to the tools of the economic theorist who proposes to understand economic phenomena through the lens of mechanism design.

Rakesh V. Vohra is the John L. and Helen Kellogg Professor of Managerial Economics and Decision Sciences at the Kellogg School of Management, Northwestern University, where he is also Director of the Center for Mathematical Studies in Economics and Management Science. He previously taught at the Fisher School of Business, Ohio State University, and is the author of *Advanced Mathematical Economics* (2005). Professor Vohra has also completed a manuscript on the principles of pricing with Lakshman Krishnamurthi, Professor of Marketing at the Kellogg School. Professor Vohra received his doctorate in mathematics from the University of Maryland.

Econometric Society Monographs

Editors:

Rosa L. Matzkin, University of California, Los Angeles
George J. Mailath, University of Pennsylvania

The Econometric Society is an international society for the advancement of economic theory in relation to statistics and mathematics. The Econometric Society Monograph series is designed to promote the publication of original research contributions of high quality in mathematical economics and theoretical and applied econometrics.

Other titles in the series:

G. S. Maddala, *Limited dependent and qualitative variables in econometrics*, 978 0 521 24143 4, 978 0 521 33825 7
Gerard Debreu, *Mathematical economics: Twenty papers of Gerard Debreu*, 978 0 521 23736 9, 978 0 521 33561 4
Jean-Michel Grandmont, *Money and value: A reconsideration of classical and neoclassical monetary economics*, 978 0 521 25141 9, 978 0 521 31364 3
Franklin M. Fisher, *Disequilibrium foundations of equilibrium economics*, 978 0 521 37856 7
Andreu Mas-Colell, *The theory of general equilibrium: A differentiable approach*, 978 0 521 26514 0, 978 0 521 38870 2
Truman F. Bewley, Editor, *Advances in econometrics – Fifth World Congress* (Volume I), 978 0 521 46726 1
Truman F. Bewley, Editor, *Advances in econometrics – Fifth World Congress* (Volume II), 978 0 521 46725 4
Hervé Moulin, *Axioms of cooperative decision making*, 978 0 521 36055 5, 978 0 521 42458 5
L. G. Godfrey, *Misspecification tests in econometrics: The Lagrange multiplier principle and other approaches*, 978 0 521 42459 2
Tony Lancaster, *The econometric analysis of transition data*, 978 0 521 4378 9
Alvin E. Roth and Marilda A. Oliviera Sotomayor, Editors, *Two-sided matching: A study in game-theoretic modeling and analysis*, 978 0 521 43788 2
Wolfgang Härdle, *Applied nonparametric regression*, 978 0 521 42950 4
Jean-Jacques Laffont, Editor, *Advances in economic theory – Sixth World Congress* (Volume I), 978 0 521 48459 6
Jean-Jacques Laffont, Editor, *Advances in economic theory – Sixth World Congress* (Volume II), 978 0 521 48460 2
Halbert White, *Estimation, inference and specification*, 978 0 521 25280 5, 978 0 521 57446 4
Christopher Sims, Editor, *Advances in econometrics – Sixth World Congress* (Volume I), 978 0 521 44459 0, 978 0 521 56610 0
Christopher Sims, Editor, *Advances in econometrics – Sixth World Congress* (Volume II), 978 0 521 44460 6, 978 0 521 56609 4
Roger Guesnerie, *A contribution to the pure theory of taxation*, 978 0 521 62956 0
David M. Kreps and Kenneth F. Wallis, Editors, *Advances in economics and econometrics – Seventh World Congress* (Volume I), 978 0 521 58983 3
David M. Kreps and Kenneth F. Wallis, Editors, *Advances in economics and econometrics – Seventh World Congress* (Volume II), 978 0 521 58982 6
David M. Kreps and Kenneth F. Wallis, Editors, *Advances in economics and econometrics – Seventh World Congress* (Volume III), 978 0 521 58013 7, 978 0 521 58981 9
Donald P. Jacobs, Ehud Kalai, and Morton I. Kamien, Editors, *Frontiers of research in economic theory: The Nancy L. Schwartz Memorial Lectures, 1983–1997*, 978 0 521 63222 5, 978 0 521 63538 7
A. Colin Cameron and Pravin K. Trivedi, *Regression analysis of count data*, 978 0 521 63201 0, 978 0 521 63567 7
Steinar Strom, Editor, *Econometrics and economic theory in the 20th century: The Ragnar Frisch Centennial Symposium*, 978 0 521 63323 9, 978 0 521 63365 9

Continued on page following the index

Mechanism Design
A Linear Programming Approach

Rakesh V. Vohra
Northwestern University, Illinois

CAMBRIDGE UNIVERSITY PRESS

CAMBRIDGE UNIVERSITY PRESS
Cambridge, New York, Melbourne, Madrid, Cape Town,
Singapore, São Paulo, Delhi, Mexico City

Cambridge University Press
32 Avenue of the Americas, New York, NY 10013-2473, USA

www.cambridge.org
Information on this title: www.cambridge.org/9780521179461

© Rakesh V. Vohra 2011

This publication is in copyright. Subject to statutory exception
and to the provisions of relevant collective licensing agreements,
no reproduction of any part may take place without the written
permission of Cambridge University Press.

First published 2011
Reprinted 2012

A catalog record for this publication is available from the British Library.

Library of Congress Cataloging in Publication Data

Vohra, Rakesh V.
Mechanism design : a linear programming approach / Rakesh V. Vohra.
  p. cm. – (Econometric Society monographs ; 47)
Includes bibliographical references and index.
ISBN 978-1-107-00436-8 (hardback) – ISBN 978-0-521-17946-1 (paperback)
1. Decision making – Linear programming. 2. Organizational behavior – Mathematical
models. 3. Machine theory. I. Title.
HD30.23.V637 2011
658.4′033–dc22   2011009042

ISBN 978-1-107-00436-8 Hardback
ISBN 978-0-521-17946-1 Paperback

Cambridge University Press has no responsibility for the persistence or accuracy of URLs
for external or third-party Internet Web sites referred to in this publication and does not
guarantee that any content on such Web sites is, or will remain, accurate or appropriate.

"Economists, unlike Lawyers, believe you have to pay people to tell the truth."

Morton Kamien

"If you were to recite even a single page in the open air, birds would fall stunned from the sky."

Clive James

Contents

1	Introduction		*page* 1
	1.1 Outline		5
2	Arrow's Theorem and Its Consequences		7
	2.1 The Integer Program		8
	2.1.1 General Domains		13
	2.2 Social Choice Functions		14
	2.2.1 Strategic Candidacy		17
	2.3 Mechanisms and Revelation		20
3	Network Flow Problem		24
	3.1 Graphs		24
	3.2 Network Flow Problem		26
	3.3 Flow Decomposition		30
	3.4 The Shortest-Path Polyhedron		33
	3.4.1 Interpreting the Dual		35
	3.4.2 Infinite Networks		36
4	Incentive Compatibility		38
	4.1 Notation		38
	4.2 Dominant Strategy Incentive Compatibility		40
	4.2.1 2-Cycle Condition		44
	4.2.2 Convex Type Spaces		45
	4.2.3 Convex Valuations		51
	4.2.4 Roberts's Theorem		54
	4.3 Revenue Equivalence		59
	4.3.1 A Demand-Rationing Example		65
	4.4 The Classical Approach		69
	4.5 Interdependent Values		73
	4.6 Bayesian Incentive Compatibility		77
5	Efficiency		79
	5.1 Vickrey-Clarke-Groves Mechanism		79
	5.2 Combinatorial Auctions		82

	5.3 The Core	87
	5.4 Ascending Auctions	89
	5.4.1 Primal-Dual Algorithm	89
	5.4.2 Incentives	98
	5.4.3 Subgradient Algorithm	99
	5.5 Gross Substitutes	105
	5.6 An Impossibility	107
	5.7 A Recipe	109
6	Revenue Maximization	110
	6.1 What Is a Solution?	112
	6.2 One-Dimensional Types	114
	6.2.1 A Formulation	117
	6.2.2 Optimal Mechanism for Sale of a Single Object	118
	6.2.3 Polyhedral Approach	121
	6.2.4 Ironing and Extreme Points	126
	6.2.5 From Expected Allocations to the Allocation Rule	129
	6.2.6 Correlated Types	130
	6.2.7 The Classical Approach	133
	6.3 Budget Constraints	135
	6.3.1 The Continuous Type Case	139
	6.4 Asymmetric Types	140
	6.4.1 Bargaining	140
	6.5 Multidimensional Types	141
	6.5.1 Wilson's Example	143
	6.5.2 Capacity-Constrained Bidders	150
7	Rationalizability	160
	7.1 The Quasilinear Case	160
	7.2 The General Case	161
References		165
Index		171

CHAPTER 1

Introduction

The 2007 Nobel Prize in economics honored a subject, mechanism design, fundamental to the study of incentives and information. Its importance is difficult to convey in a sound bite because it does not arise from a to-do list or a ten-point plan. Rather, it is an analytical framework for thinking clearly and carefully about the most fundamental of social problems: What exactly can a given institution achieve when the information necessary to make decisions is dispersed and privately held? The range of questions to which the approach can be applied is striking. To achieve a given reduction in carbon emissions, should one rely on taxes or a cap-and-trade system? Is it better to sell an Initial Public Offering (IPO) via auction or the traditional book-building approach? Would juries produce more informed decisions under a unanimity rule or that of simple majority? Mechanism design helps us understand how the answers to these questions depend on the details of the underlying environment. In turn, this helps us understand which details matter and which do not.

To get a sense of what mechanism design is, we begin with a fable, first told by the Nobelist, Ronald Coase. It involves, as all good fables do, a coal-burning locomotive and a farmer. The locomotive emits sparks that set fire to the farmer's crops. Suppose that running the locomotive yields $1,000 worth of profit for the railroad but causes $2,000 worth of crop damage. Should the railroad be made to pay for the damage it causes?

The sparks alone do no damage. One might say the farmer caused the damage by placing crops next to the railway line. It is the *juxtaposition* of sparks and crops that lead to the $2,000 worth of damage. Perhaps, then, the farmer is liable?

If you think this strange, suppose it costs the farmer $100 to ensure the safety of the crop. If we make the railroad liable for damage to the crop, what happens? The locomotive stops running.[1] Why spend $2,000 to get a return of $1,000? The farmer takes no precautions to secure the crop. As a society, we are out $1,000 – the profit the railroad would have made had it continued to run the locomotive. Now suppose we make the farmer liable. The locomotive runs.

---

[1] Assuming the absence of technology that would eliminate the sparks.

The farmer pays $100 to safeguard the crop rather than $2,000 in crop damage. On the balance, society is out only $100. If we cared about avoiding damage in the most cost-effective way possible, we should make the farmer liable.

Suppose now the railroad had access to technology that would eliminate the sparks for a price of $50. In this case, because it is cheaper for the railroad to avoid the damage, it should be made liable. If cost effectiveness is our lodestar, it puts us in a pickle, because the assignment of liability depends on the details of the particular situation. Coase's essential insight is that it does not matter how liability is assigned as long as the parties are permitted to trade the liability among themselves.

Suppose the railroad is made liable. What matters is whether or not the railroad can pay the farmer to shoulder the liability. Assume, as before, that the railroad cannot reduce the sparks emitted without shutting down the locomotive, and that the farmer can avoid the crop damage at a cost of $100. Observe that the railroad is better off paying the farmer at least $100 (and no more than $1,000) to move the crops. The farmer will also be better off. In effect, the railroad pays the farmer to assume the liability – seemingly a win-win arrangement. Thus, as long as we allow the parties concerned to trade their liabilities, the party with the least cost for avoiding the damage will shoulder the liability. In terms of economic efficiency, it does not matter who is liable for what. It matters only that the liabilities be clearly defined, easily tradeable, and enforced. It is true that the farmer and railroad care a great deal about who is held liable for what. If it is the railroad, then it must pay the farmer. If it is the farmer, the railroad pays nothing. One may prefer, for reasons quite separate from economic efficiency, to hold one party liable rather than the other. However, the outcome in terms of who does what remains the same.

Coase recognizes there are transaction costs associated with bargaining over the transfer of liabilities. Because they might overwhelm the gains to be had from bargaining, it is of fundamental importance that such costs be minimized. Nevertheless, mutually beneficial bargains fail to be struck even when transaction costs are nonexistent. Personality, ego, and history conspire to prevent agreement. These are unsatisfying explanations for *why* mutually beneficial agreements are unmade *because* they are idiosyncratic and situation specific. Mechanism design suggests another reason: The actual cost incurred by each party to avoid the damage is private information known only to themselves.

To see why, suppose the railroad incurs a cost $R of avoiding the damage whereas the farmer incurs a cost of $F to do the same. Only the railroad knows $R and only the farmer knows $F. If $F > $R, economic efficiency dictates that the railroad should incur the cost of avoiding the damage. If $F < $R, efficiency requires the farmer to shoulder the cost of avoiding the damage. In the event that $F = $R, we are indifferent as to which one incurs the cost.

Now, let us – quite arbitrarily – make the railroad liable for the damage and trust that bargaining between railroad and farmer will result in the person with the lower cost of avoiding the damage undertaking the burden to avoid the damage. If $R > $F, the railroad should pay the farmer to take on the liability.

Introduction

Furthermore, it would want to pay as little as possible – ideally no more than $F. However, the railroad does not know the magnitude of F. So, how much should it offer? The lower the offer, the less likely it will be accepted. On the other hand, if accepted, the more profitable it is to the railroad. On the flip side, the farmer has every incentive to bluff the railroad into thinking that $F is larger than it actually is so as to make a tidy profit. If the farmer is too aggressive in this regard, the railroad may walk away thinking that $R < $F. One can conceive of a variety of bargaining procedures that might mitigate these difficulties. Is there a bargaining protocol that will lead inexorably to the party with the lower cost of avoiding the damage assuming the liability?

Mechanism design approaches this question using the tools of game theory. Any such protocol can be modeled as a game that encourages each party to truthfully reveal its cost of avoiding the damage so that the correct assignment of liability can be made. The encouragement to truthfully reveal this private information is obtained with money. The monetary rewards must be generated internally, that is, there is no rich uncle waiting on the sidelines to come to the aid of either the farmer or the railroad. Thus, the question becomes a purely mathematical one: Is there a game with these properties? Myerson and Satterthwaite (1983) proved that the answer to this question was a resounding, de Gaulle – like, "NON." There is no bargaining protocol or trusted mediator that is guaranteed in all circumstances to ensure that the party with the lower cost of avoiding the damage assumes the liability. Hence, there is always the possibility that no bargain will be struck even when it is in the mutual interest of both parties to come to terms.

Thus, Coase's original observation that the assignment of liability is irrelevant because an incorrect assignment would be corrected by bargaining in the marketplace (provided transaction costs are small) is rendered false in the presence of private information. Mechanism design also suggests how liability should be assigned. Specifically, to ensure that the liability is assigned to the party with the lowest cost for avoiding the damage, the right to *avoid* the liability should be auctioned off to the highest bidder. How is this possible? Suppose our auction works as follows. We have a price clock initially set at zero. We then raise the price. At each price, we ask the bidders (railroad and farmer) whether they wish to buy the right to avoid liability at the current price. If both say "yes," continue raising the price. The instant one of them drops out, stop and sell the right to the remaining active bidder at the terminal price. Observe that the farmer will stay active as long as the current price is below $F. The railroad will stay active as long as the current price is below $R. If the farmer drops out first, it must be because $F < $R. In this case, the farmer assumes liability and the railroad pays the auctioneer $F. In short, the farmer, who had the lower cost of avoiding the damage, is saddled with the liability. If $R < $F, the reverse happens.

The fable of the railroad and the farmer involved the allocation of liability. It could just as well have involved the allocation of a property right. One is the obverse of the other. Now, the punchline. When governments create new property rights or asset classes, these should be auctioned off to ensure they are allocated in an economically efficient manner. It is exactly this reasoning that supports the allocation of spectrum rights by auction. It is exactly this reasoning that supports the allocation of permits to pollute by auction. It is exactly this reasoning that will eventually propel the Federal Aviation Authority (FAA) to use auctions to allocate arrival and departure slots at airports. Keynes said it best: "I am sure that the power of vested interests is vastly exaggerated compared with the gradual encroachment of ideas."

It is not my ambition to provide a complete account of mechanism design and its implications. My goal is more modest. It is to provide a systematic account of the underlying mathematics of the subject. The novelty lies in the approach. The emphasis is on the use of linear programming as a tool for tackling the problems of mechanism design. This is at variance with custom and practice, which have relied on calculus and the methods of analysis.[2] There are three advantages of such an approach:

1. Simplicity. Arguments based on linear programming are both elementary and transparent.
2. Unity. The machinery of linear programming provides a way to unify results from disparate areas of mechanism design.
3. Reach. It provides the ability to solve problems that appear to be beyond the reach of traditional methods.

No claim is made that the approach advocated here should supplant the traditional mathematical machinery. Rather, it is an addition to the quiver of the economic theorist who purposes to understand economic phenomena through the lens of mechanism design.

It is assumed the reader has some familiarity with game theory, the basics of linear programming, and some convex analysis. This is no more than what is expected of a first-year student in a graduate economics program. No prior knowledge of mechanism design is assumed. However, the treatment offered here will be plain and unadorned. To quote Cassel, it lacks "the corroborative detail, intended to give artistic verisimilitude to an otherwise bald and unconvincing narrative."

The point of view that animates this monograph is the product of collaborations with many individuals, including Sushil Bikhchandani, Sven de Vries, Alexey Malakhov, Rudolf Müller, Mallesh Pai, Teo Chung Piaw, Jay Sethuraman, and James Schummer. However, they are not responsible for errors of commission or omission on my part.

It was William Thomson who first suggested that I put all this down on paper. The spur was an invitation from Luca Rigotti, Pino Lopomo, and Sasa Pekec to talk about these matters at Duke University's Fuqua School. My thanks to Daniele Condorelli, Antoine Loeper, Rudolf Müller, and John Weymark,

---

[2] My colleagues refer to this as the pre-Newtonian approach to mechanism design.

who provided comments on an earlier version. George Mailath and anonymous reviewers provided invaluable suggestions on focus and intuition. An invitation from Benny Moldovanu to spend time at the Hausdorff Institute during its program on mechanism design provided valuable time to complete this project. Michael Sara was very helpful in preparing the figures. My particular thanks to Simone Galperti and Gabriel Carroll, who helped ferret out numerous blushworthy mistakes.

1.1 OUTLINE

Here is a brief outline of the other chapters.

Chapter 2

This chapter is devoted to classical social choice. There are two main results. The first is a linear inequality description of all social welfare functions that satisfy Arrows conditions. These inequalities are then employed to derive Arrow's celebrated Impossibility Theorem. The same inequalities can be employed to derive other results about social welfare functions.

The second result is a proof of the Gibbard-Satterthwaite Impossibility Theorem. A number of authors have commented on the similarities between Arrow's Theorem and the Gibbard-Satterthwaite Theorem. Reny (2001), for example, provides a unified proof of the two results. In this chapter, it is shown that the social-choice functions of the Gibbard-Satterthwaite Theorem must satisfy the same inequalities as the social welfare functions of Arrow's Theorem. Thus, impossibility in one translates immediately into impossibility in the other. This is one illustration of the unifying power of linear programming – based arguments.

The chapter closes with the revelation principle of mechanism design. Readers with prior exposure to mechanism design can skip this portion of the chapter without loss.

Chapter 3

As noted in Chapter 2, attention in the remainder of this monograph is directed to the case when utilities are quasilinear. The incentive-compatibility constraints in this case turn out to be dual to the problem of finding a shortest-path in a suitable network. This chapter introduces basic properties of the problem of finding a shortest path in a network. In fact, a problem that is more general is considered: minimum cost network flow. The analysis of this problem is not much more elaborate than needed for the shortest-path problem. Because the minimum cost flow problem arises in other economic settings, the extra generality is worth it.

Chapter 4

This chapter applies the machinery of Chapter 3 to provide a characterization of allocation rules that can be implemented in dominant as well as Bayesian incentive-compatible strategies. In addition, a general form of the Revenue Equivalence Theorem is obtained.

Chapter 5

The focus in this chapter is on mechanisms that implement the efficient outcome. The celebrated Vickrey-Clarke-Groves mechanism is derived using the results from Chapter 4. Particular attention is devoted to indirect implementations of the Vickrey-Clarke-Groves scheme in the context of combinatorial auctions. Such indirect mechanisms have an interpretation as primal-dual algorithms for an appropriate linear programming problem.

Chapter 6

This chapter applies the machinery of linear programming to the problem of optimal mechanism design. Not only are some of the classical results duplicated, but some new results are obtained, illustrating the usefulness of the approach.

Chapter 7

The subject of this brief chapter has no apparent connection to mechanism design, but it is relevant. It considers an inverse question: Given observed choices from a menu, what can we infer about preferences? Interestingly, the same mathematical structure inherent in the study of incentives appears here.

CHAPTER 2

Arrow's Theorem and Its Consequences

By custom and tradition, accounts of mechanism design begin with a genuflection in the direction of Kenneth Arrow and his (im)possibility theorem.[1] The biblical Mas-Collel, Whinston, and Green (1995), for example, introduce mechanism design by reminding the reader of Arrow's theorem, introduced some chapters earlier. Weight of history aside, there is no logical reason for this. Not disposed to being bolshy, I bow to precedent and begin with an account of Arrow's theorem. Whereas the conceptual connection to mechanism design is tenuous, the mathematical connection, as the linear programming approach reveals, is remarkably close.[2]

The environment considered involves a set Γ of alternatives (at least three). Let Σ denote the set of all strict preference orderings, that is, permutations over Γ.[3] The set of admissible preference orderings or **preference domain** for a society of n-agents will be a subset of Σ and denoted Ω. Let Ω^n be the set of all n-tuples of preferences from Ω, called **profiles**.[4] An element of Ω^n will typically be denoted as $\mathbf{P} = (\mathbf{p_1}, \mathbf{p_2}, \ldots, \mathbf{p_n})$, where $\mathbf{p_i}$ is interpreted as the preference ordering of agent i.

The objective is to identify for each profile $\mathbf{P}$ a strict preference ordering that will summarize it – a "median" or "mean" preference ordering, if you will.[5] The rule for summarizing a profile is called a social welfare function. Formally, an n-person social welfare function is a function $f : \Omega^n \mapsto \Sigma$. Thus for any $\mathbf{P} \in \Omega^n$, $f(\mathbf{P})$ is an ordering of the alternatives. We write $xf(\mathbf{P})y$ if x is ranked above y under $f(\mathbf{P})$.

There are many social welfare functions that one could imagine. One could list each one and examine its properties. To avoid this botanical exercise, Arrow suggested conditions that a social welfare function should satisfy to make it an attractive way to summarize a profile. An n-person **Arrovian social welfare**

---

[1] Arrow called it a 'possibility' theorem. His intellectual heirs added the 'im'.
[2] The treatment given here is based on Sethuraman, Teo, and Vohra (2003).
[3] The results extend easily to the case of indifference. See Sato (2006).
[4] This is sometimes called the **common preference domain**.
[5] This is not the usual motivation for the construct to be introduced, but will do for our purposes.

function (ASWF) on Ω is a function $f : \Omega^n \mapsto \Sigma$ that satisfies the following two conditions:

(1) **Unanimity:** If for $\mathbf{P} \in \Omega^n$ and some $x, y \in \Gamma$ we have $x\mathbf{p}_i y$ for all i, then $xf(\mathbf{P})y$.
(2) **Independence of Irrelevant Alternatives:** For any $x, y \in \Gamma$, suppose $\exists \mathbf{P}, \mathbf{Q} \in \Omega^n$, such that $x\mathbf{p}_i y$ if and only if $x\mathbf{q}_i y$ for $i = 1, \ldots, n$. Then $xf(\mathbf{P})y$ if and only if $xf(\mathbf{Q})y$.

The first axiom is uncontroversial. It stipulates that if all agents prefer alternative x to alternative y, then the social welfare function f must rank x above y. The second axiom states that the ranking of x and y by f is not affected by how the agents rank the other alternatives. This is not a benign axiom. Much bile and ink had been spent debating its merits. The reader interested in philosophical diversions on this matter can refer to Saari (2003).

A social welfare function that satisfies the two conditions is the *dictatorial rule*: rank the alternatives in the order of the preferences of a particular agent (the dictator). Formally, an ASWF is **dictatorial** if there is an i such that $f(\mathbf{P}) = \mathbf{p}_i$ for all $\mathbf{P} \in \Omega^n$. Clearly, the dictatorial rule is far from ideal as a rule for summarizing a profile.

An ordered pair $x, y \in \Gamma$ is called **trivial** if $x\mathbf{p}y$ for all $\mathbf{p} \in \Omega$. In view of unanimity, any ASWF must have $xf(\mathbf{P})y$ for all $\mathbf{P} \in \Omega^n$ whenever x, y is a trivial pair. If Ω consists only of trivial pairs, then distinguishing between dictatorial and nondictatorial ASWF's becomes nonsensical, so we assume that Ω contains at least one nontrivial pair. The domain Ω is **Arrovian** if it admits a nondictatorial ASWF.

The goal is to derive an integer linear programming formulation of the problem of finding an n-person ASWF. For each Ω, a set of linear inequalities is identified with the property that every feasible 0-1 solution corresponds to an n-person ASWF.[6] By examining the inequalities, we should be able to determine whether a given domain Ω is Arrovian.

2.1 THE INTEGER PROGRAM

Denote the set of all ordered pairs of alternatives by Γ^2. Let E denote the set of all agents, and S^c denote $E \setminus S$ for all $S \subseteq E$.

To construct an n-person ASWF, we exploit the independence of irrelevant alternatives, condition. The condition allows one to specify an ASWF in terms of which ordered pair of alternatives a particular subset, S, of agents is decisive over. A subset S of agents is **decisive for** x **over** y with respect to the ASWF f, if whenever all agents in S rank x over y and all agents in S^c rank y over x, the ASWF f ranks x over y.[7]

---

[6] The formulation is an extension of the **decomposability** conditions identified by Kalai and Muller (1977).
[7] In the literature, this is called *weakly* decisive.

2.1 The Integer Program

For each nontrivial element $(x, y) \in \Gamma^2$, we define a 0-1 variable as follows:

$$d_S(x, y) = \begin{cases} 1, & \text{if the subset } S \text{ of agents is decisive for } x \text{ over } y; \\ 0, & \text{otherwise.} \end{cases}$$

If $(x, y) \in \Gamma^2$ is a trivial pair, then by default we set $d_S(x, y) = 1$ for all $S \neq \emptyset$.[8]

To each ASWF f, we can associate d variables that can be determined as follows: for each $S \subseteq E$, and each nontrivial pair (x, y), pick a $\mathbf{P} \in \Omega^n$ in which agents in S rank x over y, and agents in S^c rank y over x; if $xf(\mathbf{P})y$, set $d_S(x, y) = 1$, else set $d_S(x, y) = 0$.

The remainder of this section identifies conditions satisfied by the d variables associated with an ASWF f.

Unanimity: To ensure unanimity, for all $(x, y) \in \Gamma^2$, we must have

$$d_E(x, y) = 1. \tag{2.1}$$

Independence of Irrelevant Alternatives: Consider a pair of alternatives $(x, y) \in \Gamma^2$, a $\mathbf{P} \in \Omega^n$, and let S be the set of agents that prefer x to y in $\mathbf{P}$. (Thus, each agent in S^c prefers y to x in $\mathbf{P}$.) Suppose $xf(\mathbf{P})y$. Let $\mathbf{Q}$ be any other profile such that all agents in S rank x over y and all agents in S^c rank y over x. By the independence of irrelevant alternatives, condition $xf(\mathbf{Q})y$. Hence, the set S is decisive for x over y. However, if $yf(\mathbf{P})x$, a similar argument would imply that S^c is decisive for y over x. Thus, for all S and nontrivial $(x, y) \in \Gamma^2$, we must have

$$d_S(x, y) + d_{S^c}(y, x) = 1. \tag{2.2}$$

A consequence of equations (2.1) and (2.2) is that $d_\emptyset(x, y) = 0$ for all $(x, y) \in \Gamma^2$.

Transitivity: To motivate the next class of constraints, consider majority rule. Suppose the number of agents is odd. Majority rule ranks alternative x above alternative y if a strict majority of the agents prefer x to y. Thus, majority rule can be described using the following variables:

$$d_S(x, y) = \begin{cases} 1, & \text{if } |S| > n/2, \\ 0, & \text{otherwise.} \end{cases}$$

These variables satisfy equations (2.1) and (2.2). However, if Ω admits a **Condorcet** triple (e.g., $\mathbf{p}_1, \mathbf{p}_2, \mathbf{p}_3 \in \Omega$ with $x\mathbf{p}_1 y\mathbf{p}_1 z$, $y\mathbf{p}_2 z\mathbf{p}_2 x$, and $z\mathbf{p}_3 x\mathbf{p}_3 y$), then such a rule does not always return an element of Σ for each preference profile. The reader can verify that applying majority rule to a three-agent

---

[8] To accommodate indifferences in preferences as well as the social ordering, Sato (2006) proposes a modification of the decision variables. For each $S \subseteq E$ of agents and each $(x, y) \in \Gamma^2$, $d_S(x, y) = 1$ is interpreted to mean that in any profile where all agents in S prefer x to y or are indifferent between them and all agents in $E \setminus S$ prefer y to x or are indifferent between them, then x is, socially, at least as preferred as y.

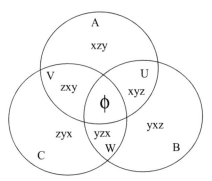

Figure 2.1 The sets and the associated orderings.

profile corresponding to a Condorcet triple does not return an ordering. The next constraint (**cycle elimination**) excludes this and similar possibilities.

For each triple x, y, z and partition of the agents in up to six sets, A, B, C, U, V, and W,

$$d_{A \cup U \cup V}(x, y) + d_{B \cup U \cup W}(y, z) + d_{C \cup V \cup W}(z, x) \leq 2, \qquad (2.3)$$

where the sets satisfy the following conditions (hereafter referred to as conditions [*]):

$A \neq \emptyset$ only if there exists $\mathbf{p} \in \Omega, x\mathbf{p}z\mathbf{p}y$,
$B \neq \emptyset$ only if there exists $\mathbf{p} \in \Omega, y\mathbf{p}x\mathbf{p}z$,
$C \neq \emptyset$ only if there exists $\mathbf{p} \in \Omega, z\mathbf{p}y\mathbf{p}x$,
$U \neq \emptyset$ only if there exists $\mathbf{p} \in \Omega, x\mathbf{p}y\mathbf{p}z$,
$V \neq \emptyset$ only if there exists $\mathbf{p} \in \Omega, z\mathbf{p}x\mathbf{p}y$,
$W \neq \emptyset$ only if there exists $\mathbf{p} \in \Omega, y\mathbf{p}z\mathbf{p}x$.

The constraint ensures that on any profile $\mathbf{P} \in \Omega^n$, the ASWF f does not produce a ranking that "cycles."

Subsequently we prove that constraints (2.1–2.3) are both necessary and sufficient for the characterization of n-person ASWF's. Before that, it will be instructive to develop a better understanding of constraints (2.3), and their relationship to the constraints identified in Kalai and Muller (1977), called **decisiveness implications**, described later in the chapter.

Suppose there are $\mathbf{p}, \mathbf{q} \in \Omega$ and three alternatives x, y and z such that $x\mathbf{p}y\mathbf{p}z$ and $y\mathbf{q}z\mathbf{q}x$. Then,

$$d_S(x, y) = 1 \Rightarrow d_S(x, z) = 1,$$

and

$$d_S(z, x) = 1 \Rightarrow d_S(y, x) = 1.$$

The first implication follows from using a profile $\mathbf{P}$ in which agents in S rank x over y over z and agents in S^c rank y over z over x. If S is decisive for x

2.1 The Integer Program

over y, then $xf(\mathbf{P})y$. By unanimity, $yf(\mathbf{P})z$. By transitivity, $xf(\mathbf{P})z$. Hence S is also decisive for x over z. The second implication follows from a similar argument.[9]

These implications can be formulated as the following two inequalities:

$$d_S(x, y) \le d_S(x, z), \quad (2.4)$$

$$d_S(z, x) \le d_S(y, x). \quad (2.5)$$

CLAIM 2.1.1 *Constraints (2.4, 2.5) are special cases of constraints (2.3).*

Proof. Let $U = S$, $W = S^c$ in constraint (2.3), with the other sets being empty. U and W can be assumed nonempty by condition (*). Constraint (2.3) reduces to $d_U(x, y) + d_{U \cup W}(y, z) + d_W(z, x) \le 2$. Because $U \cup W = E$, the above reduces to $0 \le d_S(x, y) + d_{S^c}(z, x) \le 1$, which implies $d_S(x, y) \le d_S(x, z)$ by (2.2). By interchanging the roles of S and S^c, we obtain the inequality $d_{S^c}(x, y) \le d_{S^c}(x, z)$, which is equivalent to $d_S(z, x) \le d_S(y, x)$. ∎

Suppose we know only that there is a $\mathbf{p} \in \Omega$ with $x\mathbf{p}y\mathbf{p}z$. In this instance, transitivity requires

$$d_S(x, y) = 1 \text{ and } d_S(y, z) = 1 \Rightarrow d_S(x, z) = 1,$$

and

$$d_S(z, x) = 1 \Rightarrow \text{at least one of } d_S(y, x) = 1 \text{ or } d_S(z, y) = 1.$$

These can be formulated as the following two inequalities:

$$d_S(x, y) + d_S(y, z) \le 1 + d_S(x, z), \quad (2.6)$$

$$d_S(z, y) + d_S(y, x) \ge d_S(z, x). \quad (2.7)$$

Similarly, we have:

CLAIM 2.1.2 *Constraints (2.6, 2.7) are special cases of constraints (2.3).*

Proof. Suppose $\exists \mathbf{q} \in \Omega$ with $x\mathbf{q}y\mathbf{q}z$. If there exists $\mathbf{p} \in \Omega$ with $y\mathbf{p}z\mathbf{p}x$ or $z\mathbf{p}x\mathbf{p}y$, then constraints (2.6, 2.7) are implied by constraints (2.4, 2.5), which in turn are special cases of constraint (2.3). So we may assume that there does not exist $\mathbf{p} \in \Omega$ with $y\mathbf{p}z\mathbf{p}x$ or $z\mathbf{p}x\mathbf{p}y$. If there does not exist $\mathbf{p} \in \Omega$ with $z\mathbf{p}y\mathbf{p}x$, then x, z is a trivial pair, and constraints (2.6, 2.7) are redundant. So we may assume that such a $\mathbf{p}$ exists, hence C can be chosen to be nonempty. Let

$$U = S, C = S^c$$

in constraint (2.3), with the other sets being empty. Constraint (2.3) reduces to

$$d_U(x, y) + d_U(y, z) + d_C(z, x) \le 2,$$

which is just $d_S(x, y) + d_S(y, z) + d_{S^c}(z, x) \le 2$. Thus constraint (2.6) follows as a special case of constraint (2.3). By reversing the roles of S and S^c

---

[9] Both implications were identified by Kalai and Muller (1977).

again, we can show that constraint (2.7) follows as a special case of constraint (2.3). ∎

For $n = 2$, the constraints (2.1–2.3) reduce to constraints (2.1, 2.2, 2.4–2.7). Therefore, constraint (2.3) is a generalization of the decisiveness implication conditions to $n \geq 3$.

Theorem 2.1.3 *Every feasible integer solution to (2.1)–(2.3) corresponds to an ASWF, and vice versa.*

Proof. Given an ASWF, it is easy to see that the corresponding d vector satisfies (2.1)–(2.3). Now pick any feasible solution to (2.1)–(2.3) and call it d. To prove that d gives rise to an ASWF, we show that for every profile of preferences from Ω, d generates an ordering of the alternatives. Unanimity and Independence of Irrelevant Alternatives follow automatically from the way the d_S variables are used to construct the ordering.

Suppose d does not produce an ordering of the alternatives. Then, for some profile $\mathbf{P} \in \Omega^n$, there are three alternatives x, y, and z such that d ranks x over y, y over z, and z over x. For this to happen, there must be three nonempty sets H, I, and J such that

$$d_H(x, y) = 1, \quad d_I(y, z) = 1, \quad d_J(z, x) = 1,$$

and for the profile $\mathbf{P}$, agent i ranks x over y (resp. y over z, z over x) if and only if i is in H (resp. I, J). Note that $H \cup I \cup J$ is the set of all agents, and $H \cap I \cap J = \emptyset$.

Let

$$A \leftarrow H \setminus (I \cup J), B \leftarrow I \setminus (H \cup J), C \leftarrow J \setminus (H \cup I),$$
$$U \leftarrow H \cap I, V \leftarrow H \cap J, W \leftarrow I \cap J.$$

Now A (resp. B, C, U, V, W) can only be nonempty if there exists $\mathbf{p}$ in Ω with $x\mathbf{p}z\mathbf{p}y$ (resp. $y\mathbf{p}x\mathbf{p}z$, $z\mathbf{p}y\mathbf{p}x$, $x\mathbf{p}y\mathbf{p}z$, $z\mathbf{p}x\mathbf{p}y$, $y\mathbf{p}z\mathbf{p}x$). In this case, constraint (2.3) is violated because

$$d_{A \cup U \cup V}(x, y) + d_{B \cup U \cup W}(y, z) + d_{C \cup V \cup W}(z, x)$$
$$= d_H(x, y) + d_I(y, z) + d_J(z, x) = 3. \quad ∎$$

Sato (2006) provides an interpretation of the fractional solutions of (2.1–2.3) as probabilistic ASWFs. The reader is directed there for further discussion.

Suppose $\Omega \subset \Omega'$. Then the constraints (2.1)–(2.3) corresponding to Ω are a subset of the constraints (2.1)–(2.3) corresponding to Ω'. However, one cannot infer that any ASWF for Ω' will specify an ASWF for Ω. For example, (x, y) may be a trivial pair in Ω but not in Ω'. In this case, $d_S(x, y) = 1$ for all S is a constraint in the integer program for Ω but not in the integer program for Ω'.

2.1 The Integer Program

Theorem 2.1.4 (Arrow's Impossibility Theorem) When $\Omega = \Sigma$, the 0-1 solutions to (2.1)–(2.3) correspond to dictatorial rules.

Proof. When $\Omega = \Sigma$, we know from constraints (2.4–2.5) and the existence of all possible triples that $d_S(x, y) = d_S(y, z) = d_S(z, u)$ for all alternatives x, y, z, u and for all $S \subseteq E$. We will thus write d_S in place of $d_S(x, y)$ in the rest of the proof.

We show first that for all $S \subset T$, $d_S = 1 \Rightarrow d_T = 1$. Suppose not. Let T be a set containing S such that $d_T = 0$. Constraint (2.2) implies $d_{T^c} = 1$. Choose $A = T \setminus S$, $U = T^c$ and $V = S$ in (2.3). Then $d_{A \cup U \cup V} = d_E = 1$, $d_{B \cup U \cup W} = d_{T^c} = 1$ and $d_{C \cup V \cup W} = d_S = 1$, which contradicts (2.3).

The same argument implies that $d_T = 0 \Rightarrow d_S = 0$ whenever $S \subset T$. Note also that if $d_S = d_T = 1$, then $S \cap T \neq \emptyset$, otherwise the assignment $A = (S \cup T)^c$, $U = S$, $V = T$ will violate the cycle elimination constraint. Furthermore, $d_{S \cap T} = 1$, otherwise the assignment $A = (S \cup T)^c$, $U = T \setminus S$, $V = S \setminus T$, $W = S \cap T$ will violate the cycle elimination constraint. Hence there exists a minimal set S^* with $d_{S^*} = 1$ such that all T with $d_T = 1$ contains S^*. We show that $|S^*| = 1$. If not, there will be $j \in S^*$ with $d_j = 0$, which by (2.2) implies $d_{E \setminus \{j\}} = 1$. Because $d_{S^*} = 1$ and $d_{E \setminus \{j\}} = 1$, $d_{E \setminus \{j\} \cap S^*} \equiv d_{S^* \setminus \{j\}} = 1$, contradicting the minimality of S^*. ∎

One can interpret the Impossibility Theorem as characterizing the 0-1 solutions to (2.1)–(2.3): There exists j^* such that $d_S(x, y) = 1$ for every S containing j^*. This interpretation allows one to obtain impossibility theorems for other classes of problems. See Section 2.2 for an example.

The Impossibility Theorem is the cause of much despair because it suggests the inevitability of rule by Leviathan. Rail, wail and gnash teeth. If one dislikes the conclusion, change the assumptions. Two that are prime candidates for relaxing are the independence of irrelevant alternatives assumption and the requirement that $\Omega = \Sigma$. The consequences of relaxing IIA are discussed in Saari (2001), whereas relaxing the assumption that $\Omega = \Sigma$ is discussed in Gaertner (2001).[10]

Theorem 2.1.3 can be deployed to study ASWFs that satisfy additional restrictions. For details, see Sethuraman, Teo, and Vohra (2003) and Sethuraman, Teo, and Vohra (2006).

2.1.1 General Domains

Theorem 2.1.3 can be generalized to the case in which the domain of preferences for each agent is nonidentical. In general, let D be the domain of admissible profiles over alternatives. In this case, for each set S, the d_S variables need not be defined for each pair of alternatives (x, y), if there is no profile in D in which all agents in S (resp. S^c) rank x over y (resp. y over x). Thus, d_S is defined for

---

[10] One could also relax the assumption that the ASWF produces a complete ordering. See, for example, Blair and Pollak (1982). The outcome is a form of partial dictatorship.

(x, y) only if such profiles exist. Note that $d_S(x, y)$ is well defined if and only if $d_{S^c}(y, x)$ is well defined. With this proviso, equations (2.1) and (2.2) remain valid. It suffices to modify (2.3) to the following:

Let A, B, C, U, V, and W be (possibly empty) *disjoint* sets of agents whose union includes all agents. For each such partition of the agents, and any triple (x, y, z),

$$d_{A \cup U \cup V}(x, y) + d_{B \cup U \cup W}(y, z) + d_{C \cup V \cup W}(z, x) \leq 2, \tag{2.8}$$

where the sets satisfy the following conditions:

$i \in A$ if there exists $\mathbf{p}_i$ with $x \mathbf{p}_i z \mathbf{p}_i y$,

$i \in B$ if there exists $\mathbf{p}_i$ with $y \mathbf{p}_i x \mathbf{p}_i z$,

$i \in C$ if there exists $\mathbf{p}_i$ with $z \mathbf{p}_i y \mathbf{p}_i x$,

$i \in U$ if there exists $\mathbf{p}_i$ with $x \mathbf{p}_i y \mathbf{p}_i z$,

$i \in V$ if there exists $\mathbf{p}_i$ with $z \mathbf{p}_i x \mathbf{p}_i y$,

$i \in W$ if there exists $\mathbf{p}_i$ with $y \mathbf{p}_i z \mathbf{p}_i x$.

and $(\mathbf{p}_1, \ldots, \mathbf{p}_n) \in D$.

The following theorem is immediate from our discussion.

Theorem 2.1.5 *Every feasible integer solution to (2.1), (2.2), and (2.8) corresponds to an ASWF on domain D, and vice versa.*

2.2 SOCIAL CHOICE FUNCTIONS

Arrow's Theorem is usually interpreted as asserting the impossibility of aggregating individual preferences into a coherent social preference. The desire for a coherent social ordering is to specify what outcomes a society might choose when faced with a choice among alternatives. If this is the case, why not focus on the problem of choosing one alternative given the possibly conflicting preferences of individuals who must live with the choice? In the background, another question lurks. The preferences of individuals are typically known only to themselves. What incentive does an individual have to reveal them if that information will be used to select an outcome that they deem inferior? It is these questions that motivate the next definition as well as the subject of mechanism design.

A **social choice function**, f, maps profiles of preferences into a single alternative, that is,

$$f : \Omega^n \mapsto \Gamma.$$

Notice we impose the common preference domain assumption again. One property that seems essential for a social choice function is the following.

2.2 Social Choice Functions

Pareto Optimality: Let $\mathbf{P} \in \Omega^n$ such that $x\mathbf{p}_i y$ for all $\mathbf{p}_i \in \mathbf{P}$. Then $f(\mathbf{P}) \neq y$.

The second property we require is that it should be in each agent's interest to truthfully reveal her preference ordering. To state this condition precisely, we will need some notation. Given a profile $\mathbf{P} \in \Omega$, we will write $\mathbf{p}_{-i}$ to denote the preferences of all agents other than i. Let $f : \Omega^n \mapsto \Gamma$ be a social choice function.

Strategy-Proof: For all i and any $(\mathbf{p}_i, \mathbf{p}_{-i})$ and $(\mathbf{q}_i, \mathbf{p}_{-i})$ in Ω^n, we require that

$$f(\mathbf{p}_i, \mathbf{p}_{-i})\mathbf{p}_i f(\mathbf{q}_i, \mathbf{p}_{-i}).$$

In words, no matter what other agents report, it is always in one's incentive to truthfully report one's own preference ordering. Another way to think of this condition is as follows. Imagine a game where the pure strategies of an agent are to choose an element of Ω. If agent i chooses $\mathbf{q}_i$, then the outcome of the game is $f(\mathbf{q}_1, \mathbf{q}_2, \ldots, \mathbf{q_n})$. Suppose f is strategy-proof and agent i has preferences $\mathbf{p}_i$. Then, reporting $\mathbf{p}_i$ is a dominant strategy for agent i. The power of strategy-proofness is contained in the observation that it must also apply when the roles of $\mathbf{p}_i$ and $\mathbf{q}_i$ are exchanged, that is,

$$f(\mathbf{q}_i, \mathbf{p}_{-i})\mathbf{q}_i f(\mathbf{p}_i, \mathbf{p}_{-i}).$$

Pareto Optimality is the counterpart of Unanimity for ASWFs. Strategy Proofness seems to have no counterpart. However, a third condition, the salience of which will become clear later, does.

Monotonicity: For all $x \in \Gamma$, $\mathbf{P}, \mathbf{Q} \in \Omega^n$, if $x = f(\mathbf{P})$ and $\{y : x\mathbf{p}_i y\} \subseteq \{y : x\mathbf{q}_i y\}$ $\forall i$, then $x = f(\mathbf{Q})$.

Informally, if at profile $\mathbf{P}$ the alternative x is selected by f, then in any profile where the set of alternatives ranked below x by each agent is enlarged (compared to $\mathbf{P}$), f is bound to choose x. Call a social choice function that satisfies pareto-optimality and monotonicity an Arrovian social choice function (ASCF). One example of an ASCF is the dictatorial one that, for some agent i, always selects agent i's top-ranked choice. As we show later in the chapter, every ASCF satisfies the *same* set of inequalities that an ASWF does.

Theorem 2.2.1 *Let $\Omega = \Sigma$ and $f : \Omega^n \mapsto \Gamma$ a strategy-proof social choice function. Then, f is monotone.*

Proof. Choose a profile $\mathbf{P} \in \Omega^n$ in which at least two agents have different alternatives as their top choice and suppose that $f(\mathbf{P}) = x$. Consider an agent who does not have x as her top choice and call her agent 1. Let $\mathbf{q}_1$ be a preference ordering obtained from $\mathbf{p}_1$ either by permuting the alternatives ranked below x, but keeping the rank of x invariant, or by moving some $z\mathbf{p}_1 x$ down to any position below x. By the assumption that $\Omega = \Sigma$, we know that $\mathbf{q}_1 \in \Omega$. It suffices to show that $f(\mathbf{q}_1, \mathbf{p}_{-1}) = x$ to prove monotonicity, for one can simply repeat the argument.

Let $f(\mathbf{q}_1, \mathbf{p}_{-1}) = y$. Strategy-proofness means that $y \neq z$. In addition, strategy-proofness means we cannot have $y\mathbf{p}_1 x$. Finally, by strategy-proofness we cannot have $x\mathbf{p}_1 y$. This is because $x\mathbf{p}_1 y$ implies that $x\mathbf{q}_1 y$ by the definition of $\mathbf{q}_1$ and the fact that $y \neq z$. Hence,

$$x = f(\mathbf{p}_1, \mathbf{p}_{-1}) \mathbf{q}_1 f(\mathbf{q}_1, \mathbf{p}_{-1}) = y,$$

which is a violation of strategy-proofness. Thus, $f(\mathbf{q}_1, \mathbf{p}_{-1}) = x$. ∎

Theorem 2.2.2 (Muller-Satterthwaite Theorem) *When $\Omega = \Sigma$, all ASCFs are dictatorial.*

Proof. For each subset S of agents and ordered pair of alternatives (x, y), denote by $[S, x, y]$ the set of all profiles where agents in S rank x first and y second, and agents in S^c rank y first and x second. By the hypothesis on Ω, both collections are nonempty.

For any profile $\mathbf{P} \in [S, x, y]$, it follows by pareto-optimality that $f(\mathbf{P}) \in \{x, y\}$. By monotonicity, if $f(\mathbf{P}) = x$ for one such profile $\mathbf{P}$, then $f(\mathbf{P}) = x$ for all $\mathbf{P} \in [S, x, y]$.

Suppose that for all $\mathbf{P} \in [S, x, y]$, we have $f(\mathbf{P}) \neq y$. Let $\mathbf{Q}$ be any profile (not necessarily in $[S, x, y]$) where all agents in S rank x above y, and all agents in S^c rank y above x. We show next that $f(\mathbf{Q}) \neq y$ too.

Suppose not. That is, $f(\mathbf{Q}) = y$. Let $\mathbf{Q}'$ be a profile obtained by moving x and y to the top in every agents ordering, but preserving their relative position within each ordering. So, if x was above y in the ordering under $\mathbf{Q}$, it remains so under $\mathbf{Q}'$. The same is true if y was above x. By monotonicity, $f(\mathbf{Q}') = y$. But monotonicity with respect to $\mathbf{Q}'$ and $\mathbf{P} \in [S, x, y]$ implies that $f(\mathbf{P}) = y$ is a contradiction.

Hence, if there is one profile in which all agents in S rank x above y, and all agents in S^c rank y above x, and y is not selected, then all profiles with such a property will not select y. This observation allows us to describe ASCFs using the following variables.

For each $(x, y) \in \Gamma^2$, define a 0-1 variable as follows:

- $g_S(x, y) = 1$ if all agents in S rank x above y and all agents in S^c rank y above x, then y is never selected;
- $g_S(x, y) = 0$ otherwise.

If E is the set of all agents, we require $g_E(x, y) = 1$ for all $(x, y) \in \Gamma^2$ to enforce pareto-optimality.

Consider $\mathbf{P} \in \Omega^n$, $(x, y) \in \Gamma^2$, and subset S of agents such that all agents in S prefer x to y and all agents in S^c prefer y to x. Then, $g_S(x, y) = 0$ implies that $g_{S^c}(y, x) = 1$ to ensure a selection. Hence, for all S and $(x, y) \in \Gamma^2$, we have

$$g_S(x, y) + g_{S^c}(y, x) = 1. \tag{2.9}$$

2.2 Social Choice Functions

We show that the variables g_S satisfy the cycle elimination constraints. If not, there exists a triple $\{x, y, z\}$ and set A, B, C, U, V, W such that the cycle elimination constraint is violated. Consider the profile **P** where each agent ranks the triple $\{x, y, z\}$ above the rest, and with the ordering of x, y, z depending on whether the agent is in A, B, C, U, V, or W. Because $g_{AUUUV}(x, y) = 1$, $g_{BUUUW}(y, z) = 1$, and $g_{CUVUW}(z, x) = 1$, none of the alternatives x, y, z is selected for the profile **P**. This violates pareto-optimality and thus creates a contradiction.

Hence, g_S satisfies constraints (2.1–2.3). Because $\Omega = \Sigma$, by Arrow's Impossibility Theorem, g_S corresponds to a dictatorial solution. ∎

The following is now a straightforward consequence.

Theorem 2.2.3 (Gibbard-Satterthwaite Theorem) *Let* $\Omega = \Sigma$ *and* $f : \Omega^n \mapsto \Gamma$ *a social choice function that satisfies pareto-optimality and strategy-proofness. Then, f is dictatorial.*

This is another dismal outcome. If one will not yield on pareto-optimality and strategy-proofness, one must restrict the set Ω to escape the impossibility. As in the case of ASWFs, one could identify conditions on Ω that would admit the existence of an ASCF. There is an extensive literature that does just that. This is not the path followed here. Instead, this book focuses on a special class of preferences: quasilinear. The existence of strategy-proof social choice functions is not in doubt here. Thus, we can ask for more – for example, a characterization of the class of strategy-proof social choice functions. However, before we explore that road, there is another to be examined. The Gibbard-Satterthwaite Theorem established an impossibility in the special case when agents are restricted to communicating only information about their preferences. Further, they do so in a single round. What if we allow them to communicate more than their preferences and do so using multiple rounds of communication? This is discussed in the next section. Below we give one more illustration of the use of the system (2.1)–(2.3).

2.2.1 Strategic Candidacy

One of the features of the classical social choice literature is that the alternatives are assumed as given. Yet there are many cases where the alternatives under consideration are generated by the participants. A paper that considers this issue is Dutta, Jackson, and Le Breton (2001). In that paper the alternatives correspond to the set of candidates for election. Candidates can withdraw themselves and in so doing affect the outcome of an election. Under some election rules (and an appropriate profile of preferences), a losing candidate would be better off withdrawing to change the outcome. They ask what election rules would be immune to such manipulations. Here we show how the integer programming approach can be used to derive the main result of their paper.[11]

---

[11] This is based on unpublished work with J. Sethuraman and C. P. Teo.

Let N be the set of voters and $\mathcal{K}$ the set of candidates, of which there are at least three. In this section, we use "voters" instead of "agents" and "candidates" instead of "alternatives" to be consistent with the notation of Dutta, Jackson, and Le Breton (2001). For brevity, we assume that $N \cap \mathcal{K} = \emptyset$, but it is an assumption that can be dropped. Voters have (strict) preferences over elements of $\mathcal{K}$, and the preference domain is the set of all orderings on $\mathcal{K}$ (i.e. $\Omega = \Sigma$). Candidates also have preferences over other candidates, however, each candidate must rank themselves first. There is no restriction on how they order the other candidates.

For any $\Gamma \subset \mathcal{K}$ and profile **P**, a voting rule is a function $f(\Gamma, \mathbf{P})$ that selects an element of Γ. Dutta and colleagues impose four conditions on the voting rule f:

(1) For any $\Gamma \subset \mathcal{K}$ and any two profiles **P** and **Q** that coincide on Γ, we have
$$f(\Gamma, \mathbf{P}) = f(\Gamma, \mathbf{Q}).$$

(2) If **P** and **Q** are two profiles that coincide on the set of voters N, then
$$f(\Gamma, \mathbf{P}) = f(\Gamma, \mathbf{Q}).$$

Hence, the outcome of the voting procedure does not depend on the preferences of the candidates.

(3) **Unanimity**: Let $a \in \Gamma \subset \mathcal{K}$ and consider a profile **P** where a is the top choice for all voters in the set Γ; then $f(\Gamma, \mathbf{P}) = a$.

(4) **Candidate Stability**: For every profile **P** and candidate $c \in \mathcal{K}$, we have that candidate c prefers $f(\mathcal{K}, \mathbf{P})$ to $f(\mathcal{K} \setminus c, \mathbf{P})$.

A voting rule is dictatorial if there is a voter $i \in N$ such that for all profiles **P** and all $c \in \mathcal{K}$, $f(\mathcal{K}, \mathbf{P})$ and $f(\mathcal{K} \setminus c, \mathbf{P})$ are voter i's top-ranked choices. The main result of Dutta, Jackson, and Le Breton (2001) is that the *the only voting rules that satisfy 1, 2, 3, and 4 are dictatorial*. We will derive this result using the same approach we used to derive the Muller-Satterthwaite Theorem. First, we establish an analog of the independence of irrelevant alternatives condition. This is done in the sequence of claims presented further in the chapter.

CLAIM 2.2.4 *If f is a voting rule, $f(\mathcal{K}, \mathbf{P}) = a$, and $c \neq a$, then $f(\mathcal{K} \setminus c, \mathbf{P}) = a$.*

Proof. If not, there exists a candidate c whose exit will change the outcome from a to $f(\mathcal{K} \setminus c, \mathbf{P})$. Because the outcome of the voting procedure does not depend on the preferences of the candidate, consider the situation where c ranks candidate a as the least preferred alternative. In this instance, c benefits by exiting the election. This contradicts the assumption that f is candidate-stable (cf. Condition 4). ∎

CLAIM 2.2.5 *If f is a voting rule, and **P** a profile in which every voter ranks all candidates in the set Γ above all candidates in the set Γ^c, then $f(\mathcal{K}, \mathbf{P}) \in \Gamma$.*

2.2 Social Choice Functions

Proof. The statement is true by unanimity if $|\mathcal{A}| = 1$. If $|\Gamma| = k + 1$, and $f(\mathcal{K}, \mathbf{P}) \notin \Gamma$, then by claim 2.2.4, $f(\mathcal{K} \setminus c, \mathbf{P}) \notin \mathcal{A}$ for $c \in \Gamma$. Let $\mathbf{Q}$ be a profile obtained from $\mathbf{P}$ by putting c as the least preferred candidate for all voters. Then in $\mathbf{Q}$, all voters prefer the $|k|$ candidates in $\Gamma \setminus c$ to the rest. Hence by induction, $f(\mathcal{K}, \mathbf{Q}) \in \Gamma \setminus c$. By candidate stability again, $f(\mathcal{K} \setminus c, \mathbf{Q}) \in \Gamma \setminus c$. By Condition 1, $f(\mathcal{K} \setminus c, \mathbf{P}) = f(\mathcal{K} \setminus c, \mathbf{Q})$, which is a contradiction. ∎

Let $[S, x, y]$ denote the set of profiles such that all voters in $S \subset N$ rank x and y, and all agents in S^c rank y and x as their top two alternatives.

CLAIM 2.2.6 *For all profiles* $\mathbf{P}, \mathbf{Q} \in [S, x, y]$, $f(\mathcal{K}, \mathbf{P}) = f(\mathcal{K}, \mathbf{Q})$.

Proof. We need to show that for all $\mathbf{P} \in [S, x, y]$, $f(\mathcal{K}, \mathbf{P})$ depends only on the set S.

Without loss of generality, label the candidates as $x_1, x_2, \ldots, x_n$, with $x = x_1$, $y = x_2$. Let $T(k)$ be the set of all profiles where $x_1, \ldots, x_k$ are ranked above $x_{k+1}, \ldots, x_n$. Furthermore, all voters rank x_{k+1} above x_{k+2} above x_{k+3} etc. By definition, $T(n)$ is the set of all orderings over $\mathcal{K}$.

Note that there is a unique $\mathbf{P}^* \in T(2) \cap [S, x, y]$. Suppose for all $\mathbf{P} \in T(k) \cap [S, x, y]$, $f(\mathcal{K}, \mathbf{P}) = f(\mathcal{K}, \mathbf{P}^*)$. Consider $\mathbf{Q} \in (T(k+1) \setminus T(k)) \cap [S, x, y]$ and suppose $f(\mathcal{K}, \mathbf{Q}) \neq f(\mathcal{K}, \mathbf{P}^*)$. By the previous claim, because $\mathbf{Q} \in [S, x, y]$, $f(\mathcal{K}, \mathbf{Q}) \in \{x_1, x_2\}$.

Suppose $f(\mathcal{K}, \mathbf{Q}) = x_1$, and $f(\mathcal{K}, \mathbf{P}^*) = x_2$. By candidate stability, $f(\mathcal{K} \setminus x_{k+1}, \mathbf{Q}) = x_1$, and $f(\mathcal{C} \setminus x_{k+1}, \mathbf{P}^*) = x_2$. Construct a new profile $\mathbf{R}$ from $\mathbf{Q}$ by moving the candidate x_{k+1} to the bottom of every voter's list. By Condition 1, $f(\mathcal{K} \setminus x_{k+1}, \mathbf{R}) = f(\mathcal{C} \setminus x_{k+1}, \mathbf{Q}) = x_1$. Now, $\mathbf{R} \in T(k) \cap [S, x, y]$, hence $f(\mathcal{K}, \mathbf{R}) = f(\mathcal{K}, \mathbf{P}^*)$. By claim 2.2.4, we have $f(\mathcal{K} \setminus x_{k+1}, \mathbf{R}) = f(\mathcal{K}, \mathbf{P}^*) = x_2$. This is a contradiction. ∎

Definition 2.2.7 *Let* $d_S(x, y) = 1$ *iff there is a profile* $\mathbf{P} \in [S, x, y]$ *where* $f(\mathcal{K}, \mathbf{P}) = x$.

CLAIM 2.2.8 *Constraint (2.3) is a valid inequality for the variables* d_S *defined above.*

Proof. Suppose not. Then there exist triplets x, y, z and sets A, B, C, U, V, W in condition (*) such that

$$d_{A \cup U \cup V}(x, y) = d_{B \cup U \cup W}(y, z) = d_{C \cup V \cup W}(z, x) = 1.$$

Without loss of generality, let $x_1 = x$, $x_2 = y$, $x_3 = z$, and let $\mathbf{P}$ be a profile in $T(3)$, where the preferences of the voters are given by the sets A, B, C, U, V, W. Because $f(\mathcal{K}, \mathbf{P}) \in \{x, y, z\}$, we may assume, for ease of exposition, that $f(\mathcal{K}, \mathbf{P}) = x$. By candidate stability, $f(\mathcal{K} \setminus y, \mathbf{P}) = f(\mathcal{K} \setminus z, \mathbf{P}) = x$. Let $\mathbf{P_y}$ and $\mathbf{P_z}$ be the profiles obtained from $\mathbf{P}$ by moving y and z to the bottom of all voters' list, respectively. By Condition 1, $f(\mathcal{K} \setminus y, \mathbf{P_y}) = f(\mathcal{K} \setminus z, \mathbf{P_z}) = x$. Note that for the profile $\mathbf{P_y}$, the set of voters in $C \cup V \cup W$ rank z above x, and all the other voters rank x above z. By assumption, $d_{C \cup V \cup W}(z, x) = 1$, hence $f(\mathcal{K}, \mathbf{P_y}) = z$. By candidate stability,

$f(\mathcal{K} \setminus y, \mathbf{P_y}) = f(\mathcal{K}, \mathbf{P_y}) = z$ – a contradiction. Hence constraint (2.3) must be valid for the variables d_S defined earlier. ∎

It is also easy to see that the variables d_S satisfy equations (2.1) and (2.2). We know that when all possible orderings of $\mathcal{K}$ are permitted, the only solution to (2.1), (2.2), and (2.3) is of the following kind: There is a voter i such that for all $x, y \in \mathcal{K}$ and all $S \ni i$, we have $d_S(x, y) = 1$. It remains to show that this implies dictatorship in the sense defined earlier. This can be done via induction over $T(k)$ as in claim 2.2.6. Hence the only voting rules that satisfy Conditions 1, 2, 3, and 4 are the dictatorial ones.

2.3 MECHANISMS AND REVELATION

What exactly can a given institution achieve when the information necessary to make decisions is dispersed and privately held? An answer requires a formalization of the idea of institution and the environment in which it operates.

Begin with the environment. This will consist of five items:

(1) A set N of agents.
(2) A set Γ of alternatives or outcomes.
(3) Each agent is endowed with a parameter that encodes her preferences over the elements of Γ. That parameter is called her **type**. Given a type, an agent's preferences can be represented by a utility function. We write $v(a|t)$ to denote the utility that an agent with type t assigns to $a \in \Gamma$. Further, an agent with type t (weakly) prefers $a \in \Gamma$ to $b \in \Gamma$ if and only if $v(a|t) \geq v(b|t)$.

How a type is represented depends on the application. In some cases, a type will be a preference ordering over the elements of Γ. In other applications, a type might record the monetary value assigned to various combinations of goods and services.
(4) A type space, T^i, from which permissible types for agent i are drawn. The set T^i corresponds to the set of allowable orderings over the set of outcomes.
(5) Each agent knows her own type, but not that of other agents in N. However, each agent has beliefs about the possible types of others. This is modeled by assuming that types are drawn according to a prior distribution, F, over the set of profiles of types, $\Pi_{i=1}^{n} T^i$. The distribution is assumed common knowledge among the agents. This implies that differences in information are more salient in understanding institutions than differences in beliefs. The common-knowledge assumption as a way to model incomplete information is not without its detractors. In the absence of a compelling alternative, we accept it and move on.[12]

---

[12] See Morris (1995) for a more detailed discussion of these matters.

2.3 Mechanisms and Revelation

Example 1 *Suppose a single good to be allocated among two agents. Thus, $N = \{1, 2\}$. The good can go to agent 1 or agent 2 (we exclude the possibility of not allocating the good at all). We represent the first outcome as $(1, 0)$ and the second as $(0, 1)$. Agent i assigns a monetary value $u^i \in [0, 1]$ to receiving the good and value zero to agent $j \neq i$ obtaining the good. We represent agent 1's possible type as $(u^1, 0)$ and agent 2's possible type as $(0, u^2)$. Hence, T^1 is all points in $[0, 1]$ on the abscissa and T^2 all points in $[0, 1]$ on the ordinate. Because an outcome a can be represented as a vector, and the type is a vector as well, $v(a|t) = a \cdot t$. So, for agent 1, $v((1, 0)|(u^1, 0)) = (1, 0) \cdot (u^1, 0) = u^1$. Finally, we assume that the nonzero component of each type is an independent and uniform draw from $[0, 1]$. That determines the prior.*

A mechanism $\mathcal{M}$ consists of a collection of actions, S^i, for each agent $i \in N$ and an outcome function $g : S^1 \times \ldots \times S^n \mapsto \Gamma$. A mechanism reduces an institution to a procedure for making a collective choice. The sets, $\{S^i\}_{i \in N}$, constitute the actions allowed to the agents – for example, reporting information, lobbying, and screaming. The set of actions can be very rich and complex. The setup permits all sorts of multiround games, information exchange, side bets, and contracts. The function g relates the various actions of agents to an outcome.

A mechanism $\mathcal{M}$ combined with the type space $\Pi_{i=1}^n T^i$ and prior F forms a Bayesian game of incomplete information. A strategy for agent $i \in N$ in such a game is a mapping, σ^i, from type $t^i \in T^i$ into an action $\sigma^i(t^i) \in S^i$.

Here is a mechanism for example 1. Have the agents play rock, paper, scissors. The winner gets the good. In the case of a tie, the good is assigned with equal probability to each. Hence $S^i = \{\text{rock, paper, scissors}\}$. The outcome function g is described by the rules of rock, paper, scissors. A strategy in the game induced by this mechanism is a mapping from an agent's type to one of the three actions.

In keeping with tradition, we assume the outcome of the game will be an equilibrium of that game. As the reader well knows, there are many different kinds of equilibria. So, let us focus on dominant strategy equilibrium for the moment.

Let $f : \Pi_{i=1}^n T^i \mapsto \Gamma$ be a social choice function. A mechanism $\mathcal{M} = (\Pi_{i \in N} S^i, g)$ **implements** f **in dominant strategies** if there is a dominant strategy equilibrium profile $(\sigma^1, \ldots, \sigma^n)$ of the game induced by $\mathcal{M}$ such that

$$g(\sigma^1(t^1), \ldots, \sigma^n(t^n)) = f(t^1, \ldots, t^n)$$

for all $(t^1, \ldots, t^n) \in \Pi_{i \in N} T^i$. In other words, for each profile $(t^1, \ldots, t^n) \in \Pi_{i \in N} T^i$, there is a dominant strategy equilibrium of the game induced by $\mathcal{M}$ that returns the outcome $f(t^1, \ldots, t^n)$.[13]

---

[13] Notice that the game induced by $\mathcal{M}$ may possess dominant strategy equilibrium outcomes that do not coincide with $f(t^1, \ldots, t^n)$. We could have been more demanding and asked that every equilibrium of the game induced by $\mathcal{M}$ return the outcome $f(t^1, \ldots, t^n)$. This is called strong implementation, a subject that will not concern us here. For more, see Serrano (2004).

We picked as our notion of equilibrium the dominant strategy equilibrium. We could have chosen Bayes-Nash equilibrium. In the definition of implementation, one simply replaces all appearances of the term dominant strategy by Bayes-Nash.

There is one class of mechanisms that is particularly simple. A mechanism $\mathcal{M} = (\Pi_{i \in N} S^i, g)$ is called a **direct mechanism** if $S^i = T^i$ for all $i \in N$. Hence, an action consists simply of choosing or reporting a type. A social choice function $f : \Pi_{i \in N} T^i \mapsto \Gamma$ is **truthfully implementable in dominant strategies** by a direct mechanism $\mathcal{M} = (\Pi_{i \in N} T^i, f)$ if there is a dominant strategy equilibrium $(\sigma^1(\cdot), \ldots, \sigma^n(\cdot))$ such that $\sigma^i(t^i) = t^i$ for all $i \in N$. It is easy to see that this is precisely the notion of strategy-proofness.

As an example, let the type space for each agent be Σ. Let f be a dictatorial social choice function. Note that f is strategy-proof. It is easy to see that f can be truthfully implemented in dominant strategies. The direct mechanism is simply to have each agent report their type and then apply f to the profile of reported types. Because f is strategy-proof, it is a dominant strategy for each agent to truthfully report their type. The Gibbard-Satterthwaite Theorem can now be rephrased as follows: The only pareto-optimal social choice functions that can be truthfully implemented in dominant strategies by a direct mechanism are the dictatorial ones. If we enlarge the class of mechanisms, might it be possible to implement nondictatorial pareto-optimal social choice functions? As the next result makes clear, the answer is no.

Theorem 2.3.1 Revelation Principle *Suppose $f : \Pi_{i \in N} T^i \mapsto \Gamma$ can be implemented in dominant strategies by the mechanism $\mathcal{M} = (\Pi_{i=1}^n S^i, g)$. Then, f can be truthfully implemented in dominant strategies by a direct mechanism.*

Proof. Since f can be implemented in dominant strategies by $\mathcal{M}$, there is a profile of strategies $(\sigma^1, \ldots, \sigma^n) \in \Pi_{i \in N} S^i$ that forms a dominant strategy equilibrium in the game induced by $\mathcal{M}$. Thus, for all $(t^1, \ldots, t^n) \in \Pi_{i \in N} T^i$, we have

$$g(\sigma^1(t^1), \ldots, \sigma^n(t^n)) = f(t^1, \ldots, t^n).$$

Furthermore, implementability in dominant strategies means that for all $i \in N$, $\{\rho^j : T^j \mapsto S^j\}_{j \in N}$, and $\{t^j \in T^j\}_{j \in N}$,

$$v(g(\rho^1(t^1), \ldots, \sigma^i(t^i), \ldots, \rho^n(t^n))|t^i)$$
$$\geq v(g(\rho^1(t^1), \ldots, \rho^i(t^i), \ldots, \rho^n(t^n))|t^i). \quad (2.10)$$

Consider the following direct mechanism $\mathcal{D} = (\Pi_{i \in N} T^i, h)$ where for all $(t^1, \ldots, t^n) \in \Pi_{i \in N} T^i$,

$$h(t^1, \ldots, t^n) = g(\sigma^1(t^1), \ldots, \sigma^n(t^n)) = f(t^1, \ldots, t^n).$$

It suffices to show that in the game induced by $\mathcal{D}$, it is dominant strategy for each agent i with type t^i to report t^i. Suppose not. Then there exists a profile

2.3 Mechanisms and Revelation

$(t^1, \ldots, t^n) \in \Pi_{i \in N} T^i$ and an agent $i \in N$ and a type $q \in T^i$ such that

$$v(h(q, t^{-i})|t^i) > v(h(t^i, t^{-i})|t^i)$$

if and only if

$$\Rightarrow v(g(\sigma^1(t^1), \ldots, \sigma^i(q), \ldots, \sigma^n(t^n))|t^i)$$
$$> v(g(\sigma^1(t^1), \ldots, \sigma^i(t^i), \ldots, \sigma^n(t^n)|t^i).$$

Choose any $\rho : T^i \mapsto S^i$ such that $\rho(t^i) = \sigma^i(q)$. Then, the last inequality can be written as

$$v(g(\sigma^1(t^1), \ldots, \rho(t^i), \ldots, \sigma^n(t^n))|t^i)$$
$$> v(g(\sigma^1(t^1), \ldots, \sigma^i(t^i), \ldots, \sigma^n(t^n)|t^i),$$

which contradicts (2.10).

The importance of the revelation principle is nicely expressed in the following from Myerson (2010):

> In any economic institution, individuals must be given appropriate incentives to share private information or to exert unobserved efforts. The revelation principle is a technical insight that allows us to make general statements about what allocation rules are feasible, subject to incentive constraints, in economic problems with adverse selection and moral hazard. The revelation principle tells us that, for any general coordination mechanism, any equilibrium of rational communication strategies for the economic agents can be simulated by an equivalent incentive-compatible direct-revelation mechanism, where a trustworthy mediator maximally centralizes communication and makes honesty and obedience rational equilibrium strategies for the agents.

CHAPTER 3

Network Flow Problem

The network flow problem is to determine the optimal way to route flows through a network to meet certain supply, demand, and capacity constraints. It represents a fundamental class of optimization problems with numerous applications in operations research (see Ahuja, Orlin, and Magnanti 1993). They play, as the reader will see, an important role in the analysis of mechanisms when preferences are quasilinear. Hence, the inclusion of this chapter in the book.

Network flow problems also arise in other areas of economic theory, where their appearance, unfortunately, goes unremarked. For example, the assignment model in Shapley and Shubik (1972) and the model of substitutable preferences in Milgrom (2009) are both instances of network flow problems.

3.1 GRAPHS

A **graph** is a collection of two objects. The first is a finite set $V = \{1, \ldots, n\}$ whose elements are called **nodes**. The second is a set E of (unordered) pairs of nodes called **edges**. As an example, suppose $V = \{1, 2, 3, 4\}$ and $E = \{(1, 2), (2, 3), (1, 3), (3, 4)\}$. A pictorial representation of this graph is shown in Figure 3.1.

A graph is called **complete** if E consists of every pair of nodes in V. The **end points** of an edge $e \in E$ are the two nodes i and j that define that edge. In this case, we write $e = (i, j)$. An edge of the form (i, i) is called a **loop**. In this chapter, it will be convenient to exclude such edges. However, in subsequent chapters, it will be useful for notational purposes to to allow them.

The **degree** of a node is the number of edges that contain it. In the graph of Figure 3.1, the degree of node 3 is 3 whereas the degree of node 2 is 2. A pair $i, j \in V$ is called **adjacent** if $(i, j) \in E$.

Fix a graph $G = (V, E)$ and a sequence $v^1, v^2, \ldots, v^r$ of nodes in G. A **path** is a sequence of edges $e_1, e_2, \ldots, e_{r-1}$ in E such that $e_i = (v^i, v^{i+1})$. The node v^1 is the initial node on the path, and v^r is the terminal node. An example of a path is the sequence $(1, 2), (2, 3), (3, 4)$ in Figure 3.1. A path in which no node appears more than once is called a **simple path**.

3.1 Graphs

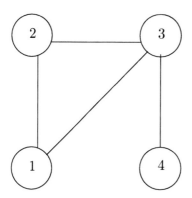

Figure 3.1 Example of a Graph.

A **cycle** is a path whose initial and terminal nodes are the same. A cycle in which all nodes except the initial and terminal nodes are distinct is called a **simple cycle**.

The edges $\{(1, 2), (2, 3), (3, 1)\}$ form a (simple) cycle in Figure 3.1. A graph G is called **connected** if there is a path in G between every pair of nodes. The graph of Figure 3.1 is connected. The graph of Figure 3.2 below is disconnected. A subset T of edges is called a **spanning tree** if the graph (V, T) is connected and acyclic. In the graph of Figure 3.1, the set $\{(1, 2), (1, 3), (3, 4)\}$ is a spanning tree.

If the edges of a graph are oriented, that is, an edge (i, j) can be traversed from i to j but not the other way around, the graph is called **directed**. If a graph is directed, the edges are called **arcs**. Formally, a directed graph consists of a set V of nodes and set A of *ordered* pairs of nodes. As an example, suppose $V = \{1, 2, 3, 4\}$ and $A = \{(1, 2), (2, 3), (3, 1), (2, 1)\}$. A pictorial representation of this graph is shown in Figure 3.3 below.

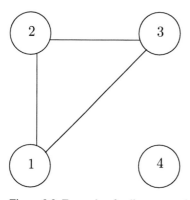

Figure 3.2 Example of a disconnected graph.

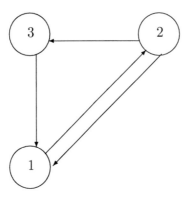

Figure 3.3 Example of a directed graph.

A path in a directed graph has the same definition as in the undirected case, except now the orientation of each arc must be respected. To emphasize this, it is common to call a path directed. In the previous example, $1 \to 3 \to 2$ would not be a directed path, but $1 \to 2 \to 3$ would be. A cycle in a directed graph is defined in the same way as in the undirected case, but again the orientation of the arcs must be respected.

A directed graph is called **strongly connected** if there is a directed path between every ordered pair of nodes. It is easy to see that this is equivalent to requiring that there be a directed cycle through every pair of nodes. If each arc in a directed graph has a number (e.g., a cost, distance, or capacity) associated with it, the directed graph is termed a **network**.

3.2 NETWORK FLOW PROBLEM

Single-commodity network flow models seek to prescribe the movement of a homogeneous good through a directed network from a set of source nodes to a set of destination nodes. Consider, for example, the network defined by $V = \{s, x, y, t\}$ and $A = \{(s, x), (s, y), (x, y), (x, t), (y, t)\}$. Figure 3.4 depicts this network. Associated with each arc of the network in Figure 3.4 is a number that can be interpreted to be the marginal cost of sending a unit flow through the relevant arc. This quantity is sometimes called an arc cost or arc length. In general, arc costs need not be positive.

To model the good's movement on a given directed network with node set V and arc set A, let x_{ij}, for each arc $(i, j) \in A$, denote the flow from i to j in an appropriately defined unit of measurement of the good on that arc. We assume conservation of flow requiring that the total flow into any node i minus the total flow $\sum_j x_{ij}$ out of that node must equal the net demand b_i at the node, that is,

$$\sum_{k:(k,i)\in A} x_{ki} - \sum_{j:(i,j)\in A} x_{ij} = b_i \; \forall i \in V.$$

3.2 Network Flow Problem

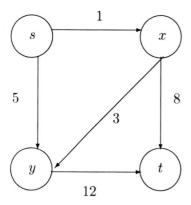

Figure 3.4 A network.

In addition to the conservation equations, network flow constraints also include lower bounds and capacities imposed upon arc flows:

$$l_{ij} \leq x_{ij} \leq u_{ij} \quad \forall (i, j) \in A.$$

These flow bounds might model physical capacities, contractual obligations, or simply operating ranges of interest. Frequently, the given lower bound capacities are all zero; $l_{ij} = -\infty$ and $u_{ij} = +\infty$ are possible as well.

For easy reference, we distinguish three types of nodes: the set $S \subseteq V$ of source, origin, or supply nodes i with $b_i < 0$; the set $T \subseteq V$ of terminal, destination, or sink nodes i with $b_i > 0$; and the set $I \subset V$ of intermediate or transshipment nodes i with $b_i = 0$. Note that, by convention, at any terminal node i, $b_i > 0$ is the net demand for flow.

Figure 3.5 illustrates a small network flow distribution system with two sources, two destinations, and one intermediate node. All lower bounds on

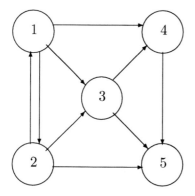

Figure 3.5 Network with two sources, two sink and one intermediate node.

flow are zero and, except for arc (3,4), are uncapacitated with upper bounds on flow equal to $+\infty$. No arc costs have been specified.

$V = \{1, 2, 3, 4, 5\}$, $A = \{(1, 2), (1, 3), (1, 4), (2, 1), (2, 3), (2, 5), (3, 4), (3, 5), (4, 5)\}$ $S = \{1, 2\}$, $I = \{3\}$ and $T = \{4, 5\}$.

The flow constraints for this model are:

$$-x_{12} - x_{13} - x_{14} + x_{21} = -2$$

$$x_{12} - x_{21} - x_{23} - x_{25} = -3$$

$$x_{13} + x_{23} - x_{34} - x_{35} = 0$$

$$x_{14} + x_{34} - x_{45} = 4$$

$$x_{25} + x_{35} + x_{45} = 1$$

$$x_{ij} \geq 0 \ \forall (i, j) \in A$$

$$x_{34} \leq 3$$

To represent network flow constraints more compactly, we use a **node-arc incidence** matrix $\mathcal{N} = \{n_{ia}\}$ with one row for each node i and one column for each arc $a \in A$. The entry n_{ia} is -1 if arc a is directed out of node i, equal to $+1$ if arc a is directed into node i, and zero otherwise. In other words, $\mathcal{N}$ is the matrix of the coefficients of the flow conservation constraints. As an example, the node-arc incidence matrix for the network of Figure 3.5 would be:

(1, 2)	(1, 3)	(1, 4)	(2, 1)	(2, 3)	(2, 5)	(3, 4)	(3, 5)	(4, 5)	arcs
−1	−1	−1	1	0	0	0	0	0	node 1
1	0	0	−1	−1	−1	0	0	0	node 2
0	1	0	0	1	0	−1	−1	0	node 3
0	0	1	0	0	0	1	0	−1	node 4
0	0	0	0	0	1	0	1	1	node 5

A network flow problem that will play an important role in what is to come is the **shortest path problem**. This problem is to find the minimum-length (sum of arc lengths) path from node s to node t in the network. In Figure 3.4, it is easy to see that the shortest path from s to t proceeds from nodes s to x to t and has a total length of 9 units. The problem can be rephrased as finding the cheapest way to route one indivisible unit of flow from s to t.

Letting l and u denote vectors of lower bounds and capacities on arc flows, and letting b denote the vector of supply/demand at the nodes, we see that in matrix form, the constraint imposed on the vector x or arc flows becomes

$$\mathcal{N}x = b, \ l \leq x \leq u.$$

Call a vector x a **feasible flow** or a feasible solution if it satisfies these flow balance and lower and upper bound conditions.

Example 2 *A special case of the network flow problem is the assignment model (see, for example, Shapley and Shubik 1972). Here the vertices of V are partitioned into two equal sized sets: workers (W) and jobs (J). Further, $b_i = -1$ for all $i \in W$, and $b_i = 1$ for all $i \in J$. In addition, no vertex in W is adjacent to another vertex in W; similarly, no vertex in J is adjacent to a vertex in J.[1] Finally, $u_{ij} = 1$, $l_{ij} = 0$ for all $i \in W$ and $j \in J$. A feasible integral flow (flows are integer valued) in this network will determine an assignment of workers to jobs such that no worker is assigned to more than one job and no job is assigned to more than one worker.*

Each variable x_{ij} appears in two flow-conservation equations, as an output from node i with a -1 coefficient and as an input to node j with a $+1$ coefficient; each column of $\mathcal{N}$ contains exactly two nonzero coefficients, a $+1$ and a -1. For now, we make three observations.

Observation 1: Summing the conservation equations eliminates all the flow variables and gives

$$0 = \sum_{j \in V} b_j \Rightarrow \sum_{j \in S} -b_j = \sum_{j \in T} b_j.$$

Consequently, total supply must equal total demand if the conservation equations $Nx = b$ are to have a feasible solution.

Observation 2: If total supply equals total demand, then summing all conservation equations gives the zero equation $0 \cdot x = 0$. Equivalently, any equation is redundant being equal to minus the sum of all other equations.

Observation 3: Let μ be a cycle in G and x^μ its incidence vector. That is, if $(i, j) \in \mu$, then $x^\mu_{ij} = 1$ and zero otherwise. If x is a feasible flow, $\epsilon \in \mathbb{R}$ and $l \leq x + \epsilon x^\mu \leq u$, then $x + \epsilon x^\mu$ is a feasible flow as well. To see why, consider any node i on the cycle μ and suppose $(p, i), (i, q) \in \mu$. We have to verify flow conservation at node i:

$$-\sum_{j:(i,j) \in A} [x_{ij} + \epsilon x^\mu_{ij}] + \sum_{k:(k,i) \in A} [x_{ki} + \epsilon x^\mu_{ki}]$$

$$= -\sum_{j:(i,j) \in A} x_{ij} - \sum_{k:(k,i) \in A} x_{ki} - \epsilon[x^\mu_{iq} - x^\mu_{pi}] = b_i,$$

where the last equation follows from the fact that $x^\mu_{iq} = x^\mu_{pi} = 1$.

Now suppose each arc is equipped with an arc cost. Let c be the vector of arc costs. Then, the cost of a feasible flow x is $\sum_{(i,j) \in A} c_{ij} x_{ij} = cx$. The network

---

[1] Such a graph is called bipartite.

flow problem is to determine a feasible flow, x, that minimizes cx:

$$\min cx$$

$$\text{s.t. } Nx = b$$

$$l \leq x \leq u.$$

One can, without loss of generality, assume that $l = 0$. This is because we can replace the variable x by $y + l$ where $0 \leq y \leq u - l$ and transform the problem into the following form:

$$\min c(y + l)$$

$$\text{s.t. } Ny = b - Nl$$

$$0 \leq y \leq u - l.$$

3.3 FLOW DECOMPOSITION

We have modeled the network flow problem in terms of arc flows x_{ij}. An alternate representation for the problem is to view the supply-and-demand requirements as being satisfied by flows along paths joining the source and sink nodes.

Let $\mathcal{P}$ denote the set of directed simple paths joining the source and sink nodes, and let $\mathcal{K}$ denote the directed simple cycles in the network. Let h_ρ be the amount of flow along the simple path $\rho \in \mathcal{P}$ and g_μ be the amount of flow around the simple cycle $\mu \in \mathcal{K}$. To convert these path and cycle flows into arc flows, we add, for each arc, the flow on all paths and cycles through that arc. Formally,

$$x_{ij} = \sum_{\rho \in \mathcal{P}} \delta_{ij}(\rho) h_\rho + \sum_{\mu \in \mathcal{K}} \delta_{ij}(\mu) g_\mu.$$

Here $\delta_{ij}(\rho) = 1$ if $(i, j) \in \rho$ and zero otherwise. Similarly $\delta_{ij}(\mu) = 1$ if $(i, j) \in \mu$ and zero otherwise. When arc flows are expressed in terms of simple path and simple cycle flows in this way, we say that the arc flows are decomposed as the sum of path and cycle flows. From now on, when a flow is decomposed, it will always be assumed that it is decomposed into flows along simple paths and simple cycles.

Of course, the arc flows x_{ij} corresponding to any given path and cycle flows need not satisfy the conservation equations $Nx = b$. They will, though, if the sum of the flows on simple paths originating from any source node equals its supply and the sum of the flows on simple paths terminating with any sink equals its demand. In this case, we say that the path and cycle flows satisfy supply-and-demand requirements.

3.3 Flow Decomposition

The following result shows that the network flow problem can be modeled in terms of arc flows or path-and-cycle flows. That is, not only can we define arc flows in terms of given path-and-cycle flows, but given an arc flow representation, we can find corresponding path and cycle flows.

Theorem 3.3.1 Flow Decomposition *An assignment of path-and-cycle flows satisfying supply-and-demand requirements corresponds to a unique assignment of arc flows satisfying the conservation equations $Nx = b$. Any arc flows fulfilling the conservation equations can be decomposed as the sum of path-and-cycle flows for some assignment of path flows h_ρ and cycle flows g_μ that satisfy supply-and-demand requirements.*

Proof. The observations preceding the theorem establish the first statement. To establish the second, we describe a procedure for converting arc flows to path-and-cycle flows. The procedure successively reduces arc flows by converting them into path-and-cycle flows. We start in step 1 with all path-and-cycle flows set equal to zero and with arc flows at the values given to us.

(1) If all arc flows as modified thus far by the procedure are equal to zero, then stop; the current path-and-cycle flows give the desired representation of arc flows. Otherwise, select some arc (i, j) with $x_{ij} > 0$, choosing, if possible, i as a source node with $b_i < 0$. Let ρ be a path with the single arc (i, j).

(2) Add to path ρ one node and arc at a time, choosing at each step any arc (k, l) directed from the last node k on path ρ and with $x_{kl} > 0$. If, at any step, node k is a terminal node with $b_k > 0$, we go to step 3; if node k appears on path ρ twice, we go to step 4. Otherwise, by conservation of flow we can find an arc (k, l) with $x_{kl} > 0$ to extend the path.

(3) ρ is a directed path from source node i to terminal node k and $x_{uv} > 0$ for every arc (u, v) on the path. Let h be the minimum of these arc flows and of $-b_i$ and b_k. Set $h_\rho = h$, redefine x_{uv} as $x_{uv} - h$ for every arc on ρ, and redefine b_i as $b_i + h$ and b_k as $b_k - h$. Return to step 1 with this modified data.

(4) The nodes and arcs from the first and second appearance of node k on ρ define a cycle μ along which all arc flows x_{uv} are positive. Let g be the minimum of the arc flows around μ. Set $g_\mu = g$, redefine x_{uv} as $x_{uv} - g$ for every arc on μ, and return to step 1 with the modified data.

If the procedure terminates in step 1, the path-and-cycle flows satisfy supply-and-demand requirements, and the arc flows are decomposed as the sum of these path-and-cycle flows. Each time step 3 or 4 is invoked, either some source or terminal node becomes a transshipment node in the modified network, or some arc flow becomes zero. Consequently, because no transshipment node ever becomes a source or terminal node, and because we never add flow to any arc, in a finite number of steps, we must terminate in step 1. ■

The theorem makes no assertion about the uniqueness of the path-and-cycle decomposition of given arc flows.[2]

Example 3 *Imagine an investor presented with a collection of certificates of deposits (CD) that are available at a certain time for a specific duration and interest rate.*[3] *If the proceeds can be reinvested, how should the investor sequence the investments to maximize their return by some given period? This problem can be formulated as a minimum cost flow problem.*

Suppose T time periods, and we are currently in time period 1. A certificate of deposit can be defined by a triple (i, j, r_{ij}) where i is the period when it is available, $j > i$ is the period when it matures, and r_{ij} is the interest rate. We can treat holding cash in a mattress as a certificate of deposit of the form $(i, i+1, 0)$.

For each time period $i \in \{1, 2, \ldots, T\}$, introduce a node. For each certificate of deposit, (i, j, r_{ij}), introduce an arc directed from i to j of cost $-\ln(1 + r_{ij})$. The capacity, u_{ij}, of the arc is set to $+\infty$. A minimum cost flow of 1 unit from node 1 to node T in this network identifies an investment strategy that maximizes the investor's returns.

Observe that the network is acyclic. So, by flow decomposition, any feasible flow can be decomposed into a collection of path flows from node 1 to node T. Consider one such path:

$$i_0 = 1 \to i_1 \to i_2 \to \cdots \to i_k = T.$$

This path is associated with investing money in the CD (i_0, i_1, r_{i_0,i_1}), taking the principal and return, and reinvesting in (i_1, i_2, r_{i_1,i_2}), and so on.

The length of this paths is

$$-\ln(1 + r_{i_0,i_1}) - \ln(1 + r_{i_1,i_2}) - \cdots - \ln(1 + r_{i_{k-1},i_k})$$
$$= \ln[\Pi_{j=0}^{k-1}(1+r_{i_j,i_{j+1}})^{-1}].$$

Thus, the absolute value of the length of a path is the reciprocal of the return implied by the investment strategy associated with this path.

Because there are no capacity constraints on the arcs, the minimum cost unit flow would identify the path with the shortest length and send the unit flow through that. The path with shortest length would correspond to the investment strategy with maximum return.

---

[2] The Flow Decomposition Theorem is a special case of Minkowski's Resolution Theorem, which says that any point in a polyhedron can be expressed as (decomposed into) a convex combination of the polyhedron's extreme points and a nonnegative linear combination of its extreme rays (see Vohra 2005). The importance of flow decomposition is that the extreme points of the shortest-path polyhedron correspond to $s - t$ paths, and extreme rays correspond to cycles in the network.

[3] A certificate of deposit is a financial instrument that pays a guaranteed interest rate provided the principal is held for a specific period.

3.4 THE SHORTEST-PATH POLYHEDRON

Let $\mathcal{N}$ be the node-arc incidence matrix of a network $G = (V, A)$ with a single source node s and sink node t. Assume there is at least one $s - t$ path. Let c_{ij} be the length of arc (i, j). Observe that the incidence vector of a $s - t$ path corresponds to a flow of a single unit along that path, and vice versa. Hence, finding a shortest $s - t$ path is equivalent to determining a minimum cost $s - t$ flow of one unit through the network. This observation allows us to define a polyhedron all of whose extreme points correspond to $s - t$ paths and every $s - t$ path is an extreme point of the polyhedron. This is the shortest-path polyhedron.

Let $b^{s,t}$ be the vector such that $b_i^{s,t} = 0$ for all $i \in V \setminus \{s, t\}$, $b_s^{s,t} = -1$ and $b_t^{s,t} = 1$. The shortest-path polyhedron is $\{x : \mathcal{N}x = b^{s,t}, x \geq 0\}$. Implicitly, $l = 0$ and $u = \infty$.

Theorem 3.4.1 *Every extreme point of $\{x : \mathcal{N}x = b^{s,t}, x \geq 0\}$ is integral.*

Proof. Let x^* be an extreme point of the polyhedron. Then, there is a c such that $x^* \in \arg\min\{cx : \mathcal{N}x = b^{s,t}, x \geq 0\}$.

Since x^* is a feasible flow, by flow decomposition, we can express x^* as the sum of positive flows around a set of simple cycles, $\mathcal{K}$, and flows along a set of simple paths from s to t, $\mathcal{P}$. For each $\mu \in \mathcal{K}$, let g_μ be the quantity of flow around the cycle μ, and for each $\rho \in \mathcal{P}$, let h_ρ be the quantity of flow along ρ. Denote by $c(\mu)$ the sum of arc lengths around the cycle and by $c(\rho)$ the sum of arc lengths along the path ρ. Hence,

$$cx^* = \sum_{\mu \in \mathcal{K}} c(\mu)g_\mu + \sum_{\rho \in \mathcal{P}} c(\rho)h_\rho.$$

First, we show that $c(\mu) \geq 0$ for all $\mu \in \mathcal{C}$. Suppose not. If $c(\mu) < 0$ for some $\mu \in \mathcal{C}$, increase the flow around μ by ϵ thus generating a new feasible flow with lower cost. This contradicts the optimality of x^*. Next, we show that if $c(\mu) > 0$ for some $\mu \in \mathcal{C}$, then $g_\mu = 0$. Suppose not. If $c(\mu) > 0$ for some $\mu \in \mathcal{K}$ with $g_\mu > 0$, reduce the flow around the cycle μ by $\epsilon > 0$. This results in a new feasible flow of lower cost, contradicting the optimality of x^*. Feasibility follows from observation 3. Hence, $cx^* = \sum_{\rho \in \mathcal{P}} c(\rho)h_\rho$.

Now suppose there exist two paths $\rho, \rho' \in \mathcal{P}$ with $h_\rho, h_{\rho'} > 0$ and $c(\rho) < c(\rho')$. Decrease the flow along ρ' by ϵ and increase the flow on ρ by the same amount. This produces a new feasible flow with lower cost – a contradiction. Hence, $c(\rho) = c(\rho')$. Let x^1 be formed by transferring ϵ units of flow from ρ' to ρ. Let x^2 be formed by transferring ϵ units of flow from ρ to ρ'. Notice that x^1 and x^2 both lie in the shortest-path polyhedron. It is easy to see that $x^* = \frac{1}{2}x^1 + \frac{1}{2}x^2$, contradicting the fact that x^* is an extreme point.

Therefore, x^* can be decomposed into a flow of one unit along a simple $s - t$ path, that is, x^* is the incidence vector of the shortest $s - t$ path. Hence, x^* is integral. ∎

COROLLARY 3.4.2 *Let $G = (V, A)$ be a network with source $s \in V$, sink $t \in V$, and arc length vector c. A shortest $s - t$ path (wrt c) exists in G iff. G contains no negative length cycles.*

Proof. Suppose first a shortest $s - t$ path in G exists. Its incidence vector x^* is an optimal extreme point of the corresponding shortest-path polyhedron. Suppose G has a negative length cycle, μ. Let x^μ be the incidence vector of this cycle. Then $x^* + \epsilon x^\mu$ for $\epsilon > 0$ is a point in the shortest-path polyhedron by observation 3 (recall the upper bound constraint on flow for each $(i, j) \in A$ is ∞). However, $cx^* + \epsilon cx^\mu < cx^*$, contradicting optimality of x^*.

Now suppose G is free of negative cycles but there is no shortest $s - t$ path. The problem of finding a shortest $s - t$ path in G can be expressed as $\min\{cx : \mathcal{N}x = b^{s,t}, x \geq 0\}$. By assumption, the program is feasible. The absence of a shortest path means this program is unbounded. Thus, for any $M < 0$, there is an x in the shortest-path polyhedron such that $cx < M$. By flow decomposition,

$$cx = \sum_{\mu \in \mathcal{K}} c(\mu) g_\mu + \sum_{\rho \in \mathcal{P}} c(\rho) h_\rho.$$

Feasibility, however, restricts h_ρ to being at most 1. Thus, the second term on the right-hand side is bounded from below. g_μ, however, is unbounded for all simple cycles μ. Thus, $c(\mu) < 0$ for at least one simple cycle μ — a contradiction. ∎

Because the shortest-path problem on G with arc length vector c can be written as $\min\{cx : \mathcal{N}x = b^{s,t}, x \geq 0\}$, it has a dual:

$$\max y_t - y_s$$

$$\text{s.t. } y\mathcal{N} \leq c.$$

The dual variables are called node potentials (and are unrestricted in sign because the primal had equality constraints). The dual has one constraint for each arc. From the structure of $\mathcal{N}$, we see that the typical dual constraint looks like:

$$y_j - y_i \leq c_{ij}$$

where $(i, j) \in A$. Thus, the set of feasible node potentials forms a lattice.

By observation 2, we know that any one of the primal constraints is redundant. Hence, we can set any one of the dual variables to zero. If we set $y_s = 0$, the dual becomes $\max\{y_t : y\mathcal{N} \leq c, y_s = 0\}$. Let y^* be an optimal solution to the dual. Therefore, by the duality theorem, y_t^* is the length of the shortest path from s to t. For any other node i, y_i^* is the length of the shortest path from s to the node i. Indeed, given any feasible dual solution y with $y_s = 0$, y_i is bounded above by the length of the shortest $s - i$ path. This can be deduced by adding up the dual constraints that correspond to arcs on the shortest path from s to i.

3.4 The Shortest-Path Polyhedron

The dual of the shortest-path problem is relevant in analyzing the feasibility of a certain class of inequalities. To describe it, let V be a finite set of indices and A a set of *ordered* pairs of elements of V.[4] Associated with each pair $(i, j) \in A$ is a number w_{ij}. The system is:

$$y_j - y_i \leq w_{ij} \quad \forall (i, j) \in A. \tag{3.1}$$

We can associate a network with (3.1) in the following way. Each element of V is a node, and to each ordered pair $(i, j) \in A$, we associate a directed arc from node i to node j. Each arc $(i, j) \in A$ is assigned a length of w_{ij}. The system (3.1) is feasible iff. the associated network contains no negative length cycle.[5] Further, it is easy to verify that the set of feasible solutions forms a lattice. The following is now straightforward.[6]

Theorem 3.4.3 *Suppose the network associated with (3.1) is strongly connected.*

(1) If the system (3.1) is feasible, one solution is to set $y_i = 0$ for an arbitrarily chosen $i \in V$ and each y_j equal to the length of the shortest path node i to node j.

(2) For any feasible solution y to (3.1) such that $y_i = 0$, each y_j is bounded above by the length of the shortest path from i to j.

(3) For any feasible solution y to (3.1) such that $y_i = 0$, each y_j is bounded below by the negative of the length of the shortest path from j to i.

The requirement that the associated network be strongly connected is needed only to ensure that there is a directed path from i to j and one from j to i. In the absence of this condition, statements (1) and (2) of Theorem 3.4.3 would make no sense.

3.4.1 Interpreting the Dual

We give two interpretations for the dual of the shortest-path polyhedron. In the first, imagine a "physical" version of the network G where each edge (i, j) is a weightless rope of length c_{ij} and each node a weightless knot. In the knot corresponding to node, t is a unit mass. Suppose we suspend the contraption of ropes and knots from a great height, h, by the node s. How far below s will the node t hang?

The configuration of ropes and knots will arrange itself to minimize the potential energy of the system. Set $y_s = 0$, which corresponds to using the location of node s as the origin. Denote by y_j the distance below s that the node j will settle.

---

[4] In fact, all of the observations to be made apply when V and A are infinite.
[5] If the direction of the inequality in (3.1) is reversed, then the system is feasible if the network has no positive length cycle.
[6] First proved in Duffin (1962).

Initially, the potential energy of the system is h (recall the contraption is of unit mass). Once all nodes except s are released, the potential energy of the system will be $h - y_t$. So, potential energy will be minimized when y_t is maximized. Now, the maximum value of y_t is constrained by the lengths of the various bars. In particular, for any i such that $(i, j) \in A$, we have $y_t \leq y_i + c_{it}$. Hence, the value of y_t that minimizes potential energy must solve

$$\max y_t$$
$$\text{s.t. } y_j - y_i \leq c_{ij} \ \forall (i, j) \in A.$$

Furthermore, when the contraption is suspended, the ropes that are taught mark out the shortest path from s to t. Equivalently, they correspond to the constraints that will bind in an optimal solution to the program above.

For the second interpretation, the nodes of the network are geographically dispersed markets into which an identical product is sold. Let y_j be the price that prevails in node (market) j. Set $y_s = 0$. We can then interpret y_j as the amount by which the price in market j exceeds the price in market s. A collection of prices exhibits an arbitrage opportunity if there is a pair of markets i and j such that $(i, j) \in A$ and $y_j > y_i + c_{ij}$. In other words, an arbitrageur could buy the product in market i for y_i, ship it to market j for a cost of c_{ij}, and sell at y_j, which would yield a profit of $y_j - (y_i + c_{ij}) > 0$. A collection of prices $\{y_j\}_{j \in V}$ is arbitrage free if

$$y_j - y_i \leq c_{ij} \ \forall (i, j) \in A.$$

Observe that these inequalities also eliminate arbitrage between markets that are not adjacent but connected by a path.

3.4.2 Infinite Networks

We claimed earlier that many of the observations derived about shortest-path lengths in networks with a finite number of vertices apply to networks with possibly uncountably may infinite vertices. The trick, if any, is to define what a shortest-path length is in a network with uncountably many vertices.

Let (V, A) be a network, with possibly uncountably many vertices. Fix an $s \in V$ and for all $t \in V$ define

$$P(t) = \inf\{\sum_{i=0}^{k} c_{t_i, t_{i+1}} | k \geq 0, s = t_0, t_1, t_2, \ldots, t_{k+1} = t \in V\}.$$

Define $P(s) = 0$. In words, $P(t)$ is the infimum of the lengths of all *finite* paths from s to t. If $P(t)$ exists, we call it the shortest-path length from s to t.

Theorem 3.4.4 *Let $G = (V, A)$ be a strongly connected network with source $s \in V$ and arc length vector c. For all $t \in V$, shortest $s - t$ path lengths (wrt c) exist in G iff. G contains no finite negative length cycles.*

3.4 The Shortest-Path Polyhedron

Proof. Suppose first that a shortest $s-t$ path lengths (wrt c) exist in G for all $t \in V$. Assume, for a contradiction, a vertex t such that there is a finite cycle C of length $\Delta < 0$. For any $-\Delta > \epsilon > 0$ we know, there is a finite $s-t$ path of length $P(t) + \epsilon$. Append to this path the finite cycle C. This yields another finite $s-t$ path of length $P(t) + \epsilon + \Delta < P(t)$, which contradicts the definition of $P(t)$.

Now assume that $G = (V, A)$ has no finite negative length cycles. Consider a pair (s, t) that lie on a cycle. Then, $P(t) + c_{t,s} \geq 0$, that is, $P(t) \geq -c_{t,s}$, so the $P(t)$ are well defined. This completes the proof. ∎

CHAPTER 4

Incentive Compatibility

The bulk of this chapter is confined to environments where types are private and independent (the independent private values case) and preferences are quasilinear. A brief discussion of the interdependent value setting is provided at the end of the chapter. Let T be a set of types (see Section 2.3). Denote by T^n the set of all n-agent profiles of types.[1] An element of T^n will usually be written as $(t^1, t^2, \ldots, t^n)$ or **t**. Let Γ be the set of outcomes. Outcomes could correspond to feasible allocations of a scarce set of resources or candidates for public office. Any direct mechanism can be decomposed into two parts: an allocation rule and a payment rule. The allocation rule determines an outcome in Γ as a function of the profile of reported types. The payment rule determines the payment each agent must make as a function of the profile of reported types. In this chapter, we show how to characterize dominant strategy as well as Bayesian incentive – compatible mechanisms in environments with quasilinear utilities and multidimensional types.

4.1 NOTATION

An **allocation rule** is a function

$$g : T^n \mapsto \Gamma.$$

For each $\alpha \in \Gamma$, let $R(g)_\alpha = \{\mathbf{t} \in T^n : g(t) = \alpha\}$. Frequently, we will supress the dependence on g and simply write R_α when there is no ambiguity. A **payment rule** is a function

$$P : T^n \mapsto \mathbb{R}^n,$$

that is, if the reported profile is $(t^1, \ldots, t^n)$, agent i makes a payment of (is taxed) $P_i(\mathbf{t})$.

The value that agent i with type $t \in T$ assigns to an allocation $\alpha \in \Gamma$ is denoted $v^i(\alpha|t)$. If agent i of type t is charged p when receiving allocation α,

---

[1] The type space does not need to be identical across agents. The assumption is for economy of notation only.

4.1 Notation

her utility is $v^i(\alpha|t) - p$. This is the assumption of quasilinearity and private values.

Here are three examples.

(1) **Homogenous, multi-item auctions with additive valuations.** Suppose we have k units of a homogenous good to allocate. The type of an agent is a vector in $\mathbb{R}^k_+$ whose j^{th} component is the marginal value for the jth unit. Each $\alpha \in \Gamma$ can be represented by an integral vector in $\mathbb{R}^k_+$ whose i^{th} component represents the quantity allocated to agent i and sum of components is k. The ith component will be denoted α_i, and $v^i(\alpha|t) = \sum_{j=1}^{\alpha_i} t_j$.

(2) **Combinatorial auctions.** We have a set M of distinct goods to allocate. The type of an agent is a vector in $\mathbb{R}^{2^{|M|}}_+$ with one component for each subset of M that corresponds to the value assigned to that subset. If the allocation α assigns the set $S \subseteq M$ to agent i with type t, then $v^i(\alpha|t) = t_S$.

(3) **Unrestricted preferences.** Following the previous example, but now we want to allow for the possibility that an agent's payoff depends not just on the goods assigned to him, but the goods assigned to other agents as well. In this case, a type would be a vector with one component for each allocation, that is, in $\mathbb{R}^{|\Gamma|}_+$.

The preceding examples are all instances of **dot-product valuations**. Specifically, for every element $\alpha \in \Gamma$, there is a vector $x(\alpha)$, and for every type $t \in T$, there is a vector y_t such that

$$v(\alpha|t) = x(\alpha) \cdot y_t.$$

When Γ is finite and $T \subseteq \mathbb{R}^{|\Gamma|}$, every valuation can be expressed as a dot-product valuation. Associate with each $\alpha \in \Gamma$ one of the unit basis vectors of $\mathbb{R}^{|\Gamma|}$ and interpret the αth component of $t \in T$, that is, t_α, as the value that type t assigns to allocation α. Then

$$v(\alpha|t) = e^\alpha \cdot t,$$

where e^α is the unit basis vector associated with α.

Suppose Γ is finite but an allocation can be any probability distribution over Γ, that is, $\Delta(\Gamma)$. Assuming risk neutrality, every valuation is a dot-product valuation. Associate each element of Γ with the corners of the simplex- $\Delta(\Gamma)$ and represent $g(\mathbf{t})$ by the relevant probability vector in $\Delta(\Gamma)$.

When Γ is infinite dimensional, a type should properly be interpreted as a function that assigns to each $\alpha \in \Gamma$ a value. Clearly, valuations in this setting will not be dot-product valuations. However, one can continue to use the dot-product notation as a convenient shorthand.

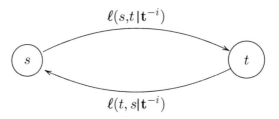

Figure 4.1 A portion of $T_g(\mathbf{t}^{-i})$.

4.2 DOMINANT STRATEGY INCENTIVE COMPATIBILITY

An allocation rule g is **implementable in dominant strategies** (IDS) if there exists a payment rule P such that for all agents i and all types $s \neq t$:

$$v^i(g(t, \mathbf{t}^{-i})|t) - P_i(t, \mathbf{t}^{-i}) \geq v^i(g(s, \mathbf{t}^{-i})|t) - P_i(s, \mathbf{t}^{-i}) \ \forall \ \mathbf{t}^{-i}. \quad (4.1)$$

An allocation rule g coupled with a payment rule P that satisfies (4.1) is called a **dominant strategy mechanism**.

The constraints (4.1) have a network interpretation. To see this, it is useful to rewrite (4.1) as

$$P_i(t, \mathbf{t}^{-i}) - P_i(s, \mathbf{t}^{-i}) \leq v^i(g(t, \mathbf{t}^{-i})|t) - v^i(g(s, \mathbf{t}^{-i})|t) \ \forall \ \mathbf{t}^{-i}. \quad (4.2)$$

They coincide with the constraints of the dual of the shortest-path problem. Given this observation, a natural step is to associate with each i and $\mathbf{t}^{-i}$ a network with one node for each type $t \in T$ and a directed arc between each ordered pair of nodes. Define the length of the arc directed from type s to type t by[2]

$$\ell(s, t | \mathbf{t}^{-i}) = v^i(g(t, \mathbf{t}^{-i})|t) - v^i(g(s, \mathbf{t}^{-i})|t).$$

Call this network $T_g(\mathbf{t}^{-i})$ (see Figure 4.1).

Corollary 3.4.2 implies that if this network has no negative length cycles, then shortest-path lengths exist. Choose any one of the nodes as a source node and set $P_i(t, \mathbf{t}^{-i})$ equal to the length of the shortest path from the source to node t.[3]

Theorem 4.2.1 (Rochet [1987]) *Let T be any type space, $n \geq 2$ be the number of agents with quasilinear utilities over a set Γ of outcomes, and $g : T^n \mapsto \Gamma$*

---

[2] Pino Lopomo suggests the phrase "from the lie to the truth," to remember the convention on how arcs are oriented.
[3] Strictly speaking, a direct application of Corollary 3.4.2 is not possible because it assumes a finite number of nodes. In the present case, the network has as many nodes as there are elements in T, which could be infinite. However, Theorem 3.4.4 allows us to do so.

4.2 Dominant Strategy Incentive Compatibility

an allocation rule. The following statements are equivalent:

(1) g is IDS.
(2) For every agent i, for every report $\mathbf{t}^{-i}$, the graph $T_g(\mathbf{t}^{-i})$ does not have a finite cycle of negative length.

Proof. Immediate from Theorem 3.4.4. ∎

In the sequel, we fix an agent i and a profile of types for the other $n-1$ agents. For this reason we suppress dependence on the index i and $\mathbf{t}^{-i}$ unless we say otherwise.

If we restrict our attention to agent i, we will always identify allocations between which agent i is indifferent. By this we can assume that for all $\alpha \neq \beta$ of Γ, there exists a type t such that $v(\alpha|t) \neq v(\beta|t)$. Furthermore, for fixed $\mathbf{t}^{-i}$, we will restrict to $\Gamma = \{\alpha | \exists t \in T \text{ such that } g(t, \mathbf{t}^{-i}) = \alpha\}$. Accordingly, from now on, $R_\alpha = \{t | g(t, \mathbf{t}^{-i}) = \alpha\}$. Under this convention, inequality (4.1) becomes

$$v(g(t)|t) - P(t) \geq v(g(s)|t) - P(s). \tag{4.3}$$

The constraints (4.2) become

$$P(t) - P(s) \leq v(g(t)|t) - v(g(s)|t) \ \forall s \in T. \tag{4.4}$$

Define the length of the arc directed from type s to type t by

$$\ell(s, t) = v(g(t)|t) - v(g(s)|t),$$

and call this network $T_g(\mathbf{t}^{-i})$.[4]

Rahman (2009) offers an interpretation of the negative cycle condition in terms of reporting strategies for the agents. This follows directly from the decision variables in the shortest-path problem on the network T_g. We describe it for the case when T is finite. Recall that we wish to know if the system

$$P(t) - P(s) \leq \ell(s, t) \ \forall s, t \in T \tag{4.5}$$

is feasible. By the Farkas lemma, system (4.5) has a solution iff. the following alternative system is infeasible:

$$\sum_{t \in T} \sum_{s \in S} \ell(s, t) x_{st} < 0 \tag{4.6}$$

$$\sum_{s \in T \setminus t} x_{st} - \sum_{s \in T \setminus t} x_{ts} = 0 \ \forall t \in T$$

$$x_{st} \geq 0 \ \forall s, t \in T.$$

Suppose system (4.5) is infeasible. Then there is a solution x^* to system (4.6). Without loss we can normalize so that $\sum_{s \in T \setminus t} x^*_{st} \leq 1$ for all $t \in T$. Set $x^*_{tt} = 1 - \sum_{s \in T \setminus t} x^*_{st}$. In this way, the x^*_{st}'s can be interpreted as probabilities.

---

[4] For technical reasons, we allow for loops. But observe that $\ell(t, t) = 0$.

In fact, given the integrality of the shortest-path polyhedron, we can assume that x^* is integral.

Interpret x^* as a reporting strategy. That is, x^*_{st} is the probability that an agent with type t should report he is type s. The first strict inequality of system 4.6 is more easily interpreted when rewritten:

$$0 < -\sum_{t \in T}\sum_{s \in S} \ell(s,t) x^*_{st} = \sum_{t \in T}\sum_{s \in S} x^*_{st}[v(g(s)|t) - v(g(t)|t)].$$

In words, the *ex ante* expected profit of following reporting strategy x^* over reporting truthfully is positive.

Formally, call a reporting strategy $\{x_{st}\}_{s,t \in T} \geq 0$ **undetectable** if

$$\sum_{s \in T} x_{st} = 1 \ \forall t \in T.$$

Call a reporting strategy $\{x_{st}\}_{s,t \in T}$ **profitable** at g if

$$\sum_{t \in T}\sum_{s \in S} x_{st}[v(g(s)|t) - v(g(t)|t)] > 0.$$

Theorem 4.2.2 *The following are equivalent:*

(1) g is IDS.
(2) There is no undetectable reporting strategy that is profitable at g.

Although the Farkas lemma applies only for finite T, the result holds more generally by appeal to duality theorems for flow problems on networks with an infinite number of vertices.[5]

An immediate consequence of Theorem 4.2.1 is what is called the **taxation principle**. If g is IDS, then we know there exist appropriate payments that implement g. Notice also, if $g(t) = g(s)$ then $P(t) = P(s)$. For this reason we can express payments as a function of the allocation rather than types, that is, $P(t) = P^\alpha$ for all $t \in R_\alpha$. Reverting to the notation involving the full profile of types, we can state the taxation principle as the following.

Lemma 4.2.3 *If g is IDS, then there exists a payment function P such that:*

$$g(\mathbf{t}) \in \arg\max_{\alpha \in \Gamma}[t^i_\alpha - P^\alpha_i(\mathbf{t}^{-i})] \ \forall i, \ \mathbf{t}^{-i}.$$

The taxation principle allows us to interpret a mechanism as offering a menu (one for each $\mathbf{t}^{-i}$) consisting of a list of allocations and corresponding prices to be paid. Specifically, if agent i chooses allocation α, she must pay $P^\alpha_i(\mathbf{t}^{-i})$.

It is sometimes convenient to work with a different but related network, called the **allocation network**. We associate with each element $\alpha \in \Gamma$ a node.

---

[5] See Rahman (2009) for a direct proof. The definitions of undetectable and profitable deviations when T is infinite dimensional remain unchanged because the relevant reporting strategies have finite support.

The length $\ell(\alpha, \beta)$, of an arc directed from allocation α to allocation β is given by

$$\ell(\alpha, \beta) = \inf_{s \in R_\beta} [v(\beta|s) - v(\alpha|s)].$$

Denote the graph by Γ_g.

When Γ is finite, we can assume that $T \subseteq \mathbb{R}^{|\Gamma|}$ and interpret the ith component of any $t \in T$ as the value that type t places on outcome $i \in \Gamma$. Therefore, if $g(t) = \alpha$, we interpret t_α to be $v(g(t)|t)$. When Γ is finite,

$$\ell(\alpha, \beta) = \inf_{s \in R_\beta} s_\beta - s_\alpha.$$

Symmetrically, we associate an arc directed from β to α with length:

$$\ell(\beta, \alpha) = \inf_{t \in R_\alpha} t_\alpha - t_\beta.$$

From Rochet's theorem (or Corollary 3.4.2), we obtain the following.

COROLLARY 4.2.4 *Let T be any type space, $n \geq 2$ be the number of agents with quasi-linear utilities over a set Γ of outcomes and $g : T^n \mapsto \Gamma$ an allocation rule. The following statements are equivalent:*

(1) g is IDS.
(2) For every agent, for every report $\mathbf{t}^{-i}$, the corresponding graph Γ_g does not have a finite cycle of negative length.

Every cycle C in T_g corresponds to a cycle C' in Γ_g. Further, the length of C is no smaller than the length of C'. If Γ is finite, the set of cycles in Γ_g is finite. Hence the set of cycles in T_g associated with a particular cycle in Γ_g form an element of a *finite* partition of the cycles in T_g.

Here is one application of Theorem 4.2.1.

Theorem 4.2.5 *Suppose $T, \Gamma \subseteq \mathbb{R}^1$, $g(t) \in \mathbb{R}^1$, $v(x|t)$ is nondecreasing in x for each fixed t and satisfies strict increasing differences (strict supermodularity). Then g is IDS iff g is nondecreasing.*

Proof. Suppose g is IDS. For a contradiction suppose g is decreasing somewhere. Then there exists $r, s \in T$ such that $r < s$ and $g(r) > g(s)$. IDS implies that

$$P(r) - P(s) \leq v(g(r)|r) - v(g(s)|r).$$

Strict increasing differences imply that the right-hand side of the above is strictly less than $v(g(r)|s) - v(g(s)|s)$. Hence

$$P(r) - P(s) < v(g(r)|s) - v(g(s)|s),$$

which violates IDS.

Now suppose that g is nondecreasing. We show that g is IDS. Increasing differences imply that for all $t > s$, we have

$$\ell(s, t) = [v(g(t)|t) - v(g(s)|t)] \geq [v(g(t)|s) - v(g(s)|s)] = -\ell(t, s).$$

Hence $\ell(s, t) + \ell(t, s) \geq 0$, that is, all cycles on pairs of nodes have nonnegative length. To show that all cycles have nonnegative length, we use induction. Suppose all cycles with k or fewer nodes have nonnegative length. For a contradiction, suppose, there is a cycle, C, of $k + 1$ nodes with negative length. Let that cycle consist of the nodes $t_1, t_2, \ldots, t_{k+1}, t_1$. Without loss, we may suppose that $t_{k+1} > t_j$ for all $j < k+1$. If we can show that $\ell(t_k, t_{k+1}) + \ell(t_{k+1}, t_1) \geq \ell(t_k, t_1)$, then the length of C is bounded below by the length of the cycle $t_1 \to t_2 \to \cdots \to t_k \to t_1$. By the induction hypothesis, this would contradict the fact that C is a negative cycle.

Suppose $\ell(t_k, t_{k+1}) + \ell(t_{k+1}, t_1) < \ell(t_k, t_1)$. Then

$$v(g(t_{k+1})|t_{k+1}) - v(g(t_k|t_{k+1}) + v(g(t_1)|t_1) - v(g(t_{k+1})|t_1)$$
$$< v(g(t_1)|t_1) - v(g(t_k)|t_1). \tag{a}$$

The right-hand side of inequality (a) can be written as

$$v(g(t_1)|t_1) - v(g(t_{k+1})|t_1) + v(g(t_{k+1})|t_1) - v(g(t_k)|t_1). \tag{b}$$

Substituting (b) into right hand side of (a) we get

$$v(g(t_{k+1})|t_{k+1}) - v(g(t_k)|t_{k+1}) < v(g(t_{k+1})|t_1) - v(g(t_k)|t_1),$$

which cannot be by increasing differences. ∎

The proof is instructive in that it suggests that sometimes it is sufficient to verify nonnegativity of cycles on pairs of nodes. This is what we investigate next.

4.2.1 2-Cycle Condition

Reversing the roles of t and s in (4.3) implies

$$v(g(s)|s) - P(s) \geq v(g(t)|s) - P(t). \tag{4.7}$$

Adding (4.3) to (4.7) yields

$$v(g(t)|t) + v(g(s)|s) \geq v(g(s)|t) + v(g(t)|s).$$

Rewriting:

$$v(g(t)|t) - v(g(s)|t) \geq -[v(g(s)|s) - v(g(t)|s)]. \tag{4.8}$$

We call (4.8) **a 2-cycle inequality**. The 2-cycle inequality in the network Γ_g for a pair $\alpha, \beta \in \Gamma$ is

$$\ell(\alpha, \beta) + \ell(\beta, \alpha) \geq 0. \tag{4.9}$$

4.2 Dominant Strategy Incentive Compatibility

An allocation rule that satisfies the 2-cycle inequality for every pair $s, t \in T$ is said to satisfy the 2-cycle condition. This condition is also called monotonicity of the allocation rule.[6] To see why monotonicity is appropriate, assume the valuations can be written in dot-product form. Thus, $g(t)$ i and t are vectors of the same dimension and $v(g(t)|t) = t \cdot g(t)$. The 2-cycle condition becomes

$$(t - s) \cdot (g(t) - g(s)) \geq 0 \;\; \forall t, s \in T.$$

However, because the author is pig-headed, the more evocative term 2-cycle is used.

That the 2-cycle condition is a necessary condition for dominant strategy incentive compatibility. In certain circumstances, it is a sufficient condition. One obvious case is when $|\Gamma| = 2$.

4.2.2 Convex Type Spaces

When T is convex and $|\Gamma|$ finite, the 2-cycle condition is sufficient for IDS. This was proved by Saks and Yu (2005). The proof given here is new.[7]

Recall that when Γ is finite, we can assume that $T \subseteq \mathbb{R}^{|\Gamma|}$ and interpret the ith component of any $t \in T$ as the value that type t places on outcome $i \in \Gamma$. Therefore, if $g(t) = \alpha$ then $v(g(t)|t) = t_\alpha$. If (4.9) holds, the set R_α is contained in a polyhedron Q_α for all $\alpha \in \Gamma$, and these polyhedra can be chosen such that the intersection of T with the interior, $I(Q_\alpha)$, of Q_α is contained in R_α. To describe Q_α, observe that for $t \in R_\alpha$ we have

$$t_\alpha - t_\beta \geq \inf_{x \in R_\alpha} [x_\alpha - x_\beta] = \ell(\beta, \alpha). \tag{4.10}$$

Thus R_α is a subset of

$$Q_\alpha = \{x \in \mathbb{R}^{|\Gamma|} : x_\alpha - x_\beta \geq \ell(\beta, \alpha) \;\; \forall \beta \neq \alpha\}.$$

Now assume $I(Q_\alpha) \neq \emptyset$ and consider a $t \in T \cap I(Q_\alpha)$. We show that $t \in R_\alpha$. Observe that for all $\beta \neq \alpha$, $t \notin R_\beta$. Indeed, otherwise we get the contradiction[8]

$$t_\beta - t_\alpha \leq \ell(\alpha, \beta) < t_\beta - t_\alpha.$$

The last inequality follows from (4.9).

---

[6] Bikhchandani, Chatterji, Lavi, Mu'alem, Nisan, and Sen (2006) called it **weak monotonicity**. It is the natural extension of monotonicity from one dimension to multiple dimensions, and so the qualifier weak should be omitted.

[7] Special cases of it had been identified earlier – see, for example, Bikhchandani, Chatterji, Lavi, Mu'alem, Nisan, and Sen (2006). Other proofs of this theorem can be found in Ashlagi, Braverman, Hassidim, and Monderer (2009) and Archer and Kleinberg (2008). It was implicit in Jehiel, Moldovanu, and Stachetti (1999).

[8] We make use of $v(\alpha|\cdot) \neq v(\beta|\cdot)$ for $\alpha \neq \beta$. This ensures that inequality (4.10) has nonzero left-hand side, and thus every point in the interior of Q_α satisfies (4.10) with strict inequality.

Notice, there is a one-to-one correspondence between the constraints of these polyhedra and arcs of Γ_g. Specifically, the constraint $x_\beta - x_\alpha \geq \ell(\alpha, \beta)$ corresponds to the arc (α, β).

The next lemma is based on ideas used in the proof of Theorem 4.2.5.

Lemma 4.2.6 *Suppose Γ is finite and $T \subseteq \mathbb{R}^{|\Gamma|}$ is convex. Then, for any $s, t \in T$ and $\epsilon > 0$, there is a sequence $r^1, r^2, \ldots, r^k \in T$ such that*

$$\ell(s, r^1) + \ell(r^1, r^2) + \cdots + \ell(r^k, t) + \ell(t, r^k) + \cdots + \ell(r^1, s) \leq \epsilon.$$

Proof. Let $d = t - s$ and choose an integer k sufficiently large. Let $r^j = s + (\frac{j}{k+1})d$ and $\mu_j = g(r_j)$ for $j = 1, \ldots, k$. Set $r^0 = s$, $\mu_0 = g(s)$, $r^{k+1} = t$ and $\mu_{k+1} = g(t)$. Now

$$\ell(r^j, r^{j+1}) = (s_{\mu_{j+1}} + \frac{j+1}{k+1} d_{\mu_{j+1}}) - (s_{\mu_j} + \frac{j+1}{k+1} d_{\mu_j}) \quad \forall r = 1, \ldots, k$$

and

$$\ell(r^{j+1}, r^j) = (s_{\mu_j} + \frac{j}{k+1} d_{\mu_j}) - (s_{\mu_{j+1}} + \frac{j}{k+1} d_{\mu_{j+1}}) \quad \forall r = 1, \ldots, k.$$

Hence

$$\ell(r^j, r^{j+1}) + \ell(r^{j+1}, r^j) = (\frac{1}{k+1})(d_{\mu_{j+1}} - d_{\mu_j}).$$

Therefore

$$\ell(s, r^1) + \ell(r^1, r^2) + \cdots + \ell(r^k, t) + \ell(t, r^k) + \cdots + \ell(r^1, s)$$
$$= (\frac{1}{k+1})(d_{\mu_{k+1}} - d_{\mu_0}) \leq \epsilon$$

for k sufficiently large. ∎

Given any two points $s, t \in T$, denote by $[s, t]_\epsilon$ the sequence of points identified in Lemma 4.2.6 when traversed from s to t.

Theorem 4.2.7 (Saks and Yu [2005]) *Suppose Γ is finite, $T \subseteq \mathbb{R}^{|\Gamma|}$ is closed and convex. Then, g is IDS iff.*

$$v(g(t)|t) - v(g(s)|t) \geq -[v(g(s)|s) - v(g(t)|s)] \quad \forall t, s \in T.$$

Proof. We confine ourselves to a proof that the 2-cycle condition implies that g is IDS. Let $\hat{C} = \{r^1 \to r^2 \to \cdots \to r^p \to r^1\}$ be a finite cycle in T_g and denote its length by $\ell(\hat{C})$. Suppose, for a contradiction, that $\ell(\hat{C}) < 0$. We show first that we may suppose that $\{r^1, r^2, \ldots, r^p\}$ contains at most three extreme points. Suppose otherwise, then $\hat{C}$ must have at least four extreme points. Pick two extreme points that are not adjacent on $\hat{C}$. Without loss, call them r^1 and r^k. Now 'split' the cycle $\hat{C}$ into two cycles:

$$C^1 : r^1 \to r^2 \to \cdots \to r^k \to [r^k, r^1]_\epsilon \to r^1$$

$$C^2 : r^k \to r^{k+1} \to \cdots \to r^1 \to [r^1, r^k]_\epsilon \to r^k$$

4.2 Dominant Strategy Incentive Compatibility

By Lemma 4.2.6,

$$\ell(C^1) + \ell(C^2) \leq \ell(\hat{C}) + \epsilon < 0$$

for ϵ sufficiently small. Without loss, we may suppose that $\ell(C^1) < 0$. Notice that C^1 will have one less extreme point than C. Repeating this argument, we conclude the existence of a negative cycle with at most three extreme points.

Suppose the convex hull of $\hat{C}$ is a triangle and let H be the two-dimensional hyperplane that contains $\hat{C}$. From now on, we restrict attention to $T' = T \cap H$. Let $C = \{t^1, t^2, \ldots, t^p\} \subset T'$ be a finite cycle in T'_g and

$$h(C) = \min\{area\{x^1, x^2, \ldots, x^p\} : x^j \in T' \cap Q_{g(t^j)} \; \forall j\}.$$

Now $h(C)$ is well defined because we are optimizing a continuous function over a compact set.

Among all negative cycles C in T', choose one with the smallest $h(C)$. Call it $Y = \{y^1, y^2, \ldots, y^p\}$. Even though the set of possible cycles in T'_g is uncountable, this is well defined because Γ is finite. Recall that each cycle in T_g, corresponds to a cycle in Γ_g, and there are only a finite number of cycles in Γ_g. Further, if C and C' are two cycles in T'_g that correspond to the same cycle in Γ_g, then $h(C) = h(C')$.

Let

$$\{t^1, \ldots, t^p\} \in \arg\min\{area\{x^1, x^2, \ldots, x^p\} : x^j \in T' \cap Q_{g(y^j)} \; \forall j\}$$

and $c = \frac{\sum_{j=1}^{p} t^j}{p}$. Note that c is in the convex hull of $\{t^1, \ldots, t^p\}$. Denote by Y' the cycle $t^1 \to t^2 \to \cdots \to t^p \to t^1$. Because Y and Y' correspond to the same cycle in Γ_g, it follows that $\ell(Y) = \ell(Y') < 0$ and $h(Y) = h(Y')$.

Since $h(Y') > 0$, $area(Y') > 0$. Denote by K^i for $i = 1, \ldots, p-1$ the cycle

$$t^i \to t^{i+1} \to [t^{i+1}, c]_\epsilon \to [c, t^i]_\epsilon.$$

K^p is the cycle $t^p \to t^1 \to [t^1, c]_\epsilon \to [c, t^p]_\epsilon$. By Lemma 4.2.6,

$$\ell(K^1) + \cdots + \ell(K^p) = \ell(Y') + p\epsilon < 0$$

for ϵ sufficiently small. Hence, $\ell(K^i) < 0$ for at least one i. However, $area(K^i) < area(Y')$ for all i – a contradiction.

Therefore, we may assume that $h(Y) = 0$, that is, $area(Y) = 0$. Thus, Y is one-dimensional (if zero-dimensional, the proof is complete). For economy of exposition only (considering that the proof will mimic that of Theorem 4.2.5), suppose that Y is the cycle $r \to s \to t \to r$. Because Y is one-dimensional, we may suppose that $s = \mu r + (1 - \mu)t$ for some $\mu \in (0, 1)$. Now, $\ell(Y) < 0$

means that

$$0 > s_{g(s)} - s_{g(r)} + t_{g(t)} - t_{g(s)} + r_{g(r)} - r_{g(t)}$$
$$\geq s_{g(s)} - s_{g(r)} + s_{g(t)} - s_{g(s)} + r_{g(r)} - r_{g(t)}$$
$$= -s_{g(r)} + s_{g(t)} + r_{g(r)} - r_{g(t)}$$
$$= \mu t_{g(t)} + (1-\mu) r_{g(t)} - \mu t_{g(r)} - (1-\mu) r_{g(r)} + r_{g(r)} - r_{g(t)}$$
$$= \mu [t_{g(t)} - t_{g(r)}] + \mu [r_{g(r)} - r_{g(t)}] \geq 0,$$

which is a contradiction. ∎

Examination of the proof given suggests an alternative formulation whose proof is straightforward.

Theorem 4.2.8 *Suppose Γ is finite, $T \subseteq \mathbb{R}^{|\Gamma|}$ is closed and convex. Suppose g is IDS when restricted to any one-dimensional subset of T. Then, g is IDS.*

Theorem 4.2.8 implies that verifying whether an allocation rule g is IDS can be reduced to verifying whether g is IDS when the type space is one-dimensional.

Convexity of the type space is essential to the validity of the conclusion of the Theorem 4.2.7 and 4.2.8. They fail, for example, when the type space is discrete. However, there are nonconvex type spaces where the 2-cycle conditions are sufficient for IDS. An example is given below.[9] The requirement that T be closed in Theorem 4.2.7 and 4.2.8 is for convenience only and can be relaxed.

Example 4 Consider the allocation of up to two indivisible objects $\{a, b\}$ to a single agent. Thus, $\Gamma = \{\emptyset, \{a\}, \{b\}, \{a, b\}\}$. The type space $\hat{T} \subseteq \mathbb{R}^4$ consists of nonnegative vectors $(0, t_a, t_b, t_{a,b})$ such that $t_{a,b} = \max\{t_a, t_b\}$. Notice that $\hat{T}$ is not convex. Let $f : \hat{T} \mapsto \Gamma$ be an allocation rule that satisfies the 2-cycle condition. We will show that f must be IDS.

Let $R_0 = \{s \in \hat{T} : f(s) = \emptyset\}$, $R_a = \{s \in \hat{T} : f(s) = a\}$, $R_b = \{s \in \hat{T} : f(s) = b\}$, and $R_{ab} = \{s \in \hat{T} : f(s) = \{a, b\}\}$. Subdivide R_{ab} into sets $H_a = \{s \in R_{ab} : s_a = s_{ab}\}$ and $H_b = R_{ab} \setminus H_a$. Let $T \subset \mathbb{R}_+^3$ be the type space consisting of $(0, s_a, s_b)$, that is, a projection from $\hat{T}$ onto to the first three components. Notice that T is convex.

Define an allocation rule g on T as follows:

(1) If $(0, s_a, s_b, s_{ab}) \in R_0$ then $g(0, s_a, s_b) = \emptyset$.
(2) If $(0, s_a, s_b, s_{ab}) \in R_a \cup H_a$ then $g(0, s_a, s_b) = a$.
(3) If $(0, s_a, s_b, s_{ab}) \in R_b \cup H_b$, then $g(0, s_a, s_b) = b$.

It is easy to check that if f satisfies the 2-cycle condition, so does g. Hence, by Theorem 4.2.7, g must be IDS.

If $\hat{t}$ and $\hat{s}$ are two types in $\hat{T}$, then $\ell_f(\hat{t}, \hat{s}) = v(f(\hat{s})|\hat{s}) - v(f(\hat{t})|\hat{s})$. Denote by t and s the projections of type $\hat{t}$ and $\hat{s}$ onto the T space. Then $\ell_g(t, s) = \ell_f(\hat{t}, \hat{s})$ when $\hat{t}, \hat{s} \in R_0 \cup R_a \cup R_b$.

---

[9] However, as shown in Ashlagi, Braverman, Hassidim, and Monderer (2009), for every nonconvex type space, there is a randomized allocation rule that is monotone but not IDS.

4.2 Dominant Strategy Incentive Compatibility

To show that f is IDS, suppose otherwise. Then, there must be a negative cycle in Γ_f. If this cycle uses only types in $R_0 \cup R_a \cup R_b$, we get a contradiction by the observation above. So, the cycle must involve one type in R_{ab} – call it $\hat{y}$. Let $\hat{x} \in \hat{T}$ and $\hat{z} \in \hat{T}$ be the types immediately preceding and succeeding type $\hat{y}$ on this cycle. If $\hat{x}$ or $\hat{z} \in R_{ab}$, we could delete them without increasing the length of the cycle. To obtain a contradiction, it will suffice to show that

$$\ell_f(\hat{x}, \hat{y}) + \ell_f(\hat{y}, \hat{z}) \geq \ell_g(x, y) + \ell_g(y, z).$$

Suppose first that $\hat{x} \in R_b$ (a similar argument applies when $\hat{x} \in R_a$). Then $\ell_f(\hat{x}, \hat{y}) = v(\{a,b\}|\hat{y}) - v(b|\hat{y}) = \hat{y}_{ab} - \hat{y}_b$. If $\hat{y} \in H_a$, then $\ell_f(\hat{x}, \hat{y}) = \hat{y}_a - \hat{y}_b$, and if $\hat{y} \in H_b$, then $\ell_f(\hat{x}, \hat{y}) = 0$. Observe that

$$\ell_g(x, y) = v(g(y)|y) - v(g(x)|y) = v(g(y)|y) - v(b|y).$$

If $\hat{y} \in H_a$, that is, $\hat{y}_a \geq \hat{y}_b$, $g(y) = a$. Hence,

$$\ell_g(x, y) = v(a|y) - v(b|y) = 0 \leq \hat{y}_a - \hat{y}_b = \ell_f(x, y).$$

If $\hat{y} \in H_b$, then $\hat{y}_b \geq \hat{y}_a$ and $g(y) = b$. Hence,

$$\ell_g(x, y) = v(b|y) - v(b|y) = \hat{y}_b - \hat{y}_b = 0.$$

Now turn to $\hat{z}$. Suppose that $\hat{z} \in R_b$ (a similar argument will apply when $\hat{z} \in R_a$). Then $\ell_f(\hat{y}, \hat{z}) = v(f(\hat{z})|\hat{z}) - v(f(\hat{y})|\hat{z}) = v(b|\hat{z}) - v(\{a,b\}|\hat{z}) = \hat{z}_b - \hat{z}_{ab}$. Now

$$\ell_g(y, z) = v(g(z)|z) - v(g(y)|z) = v(b|z) - v(g(y)|z)$$
$$= \hat{z}_b - v(g(y)|z).$$

If $\hat{y} \in H_a$, then, $g(y) = a$. Hence,

$$\ell_g(y, z) = \hat{z}_b - v(a|z) = \hat{z}_b - \hat{z}_a.$$

If $\hat{y} \in H_b$, then $g(y) = b$. Hence,

$$\ell_g(y, z) = \hat{z}_b - \hat{z}_b.$$

We now show that

$$\ell_f(\hat{x}, \hat{y}) + \ell_f(\hat{y}, \hat{z}) \geq \ell_g(x, y) + \ell_g(y, z).$$

Suppose that $\hat{x} \in R_b$, $\hat{z} \in R_b$, $\hat{y} \in H_a$, and $\hat{z}_a \geq \hat{z}_b$. Then

$$\ell_f(\hat{x}, \hat{y}) + \ell_f(\hat{y}, \hat{z}) = \hat{y}_a - \hat{y}_b + \hat{z}_b - \hat{z}_{ab} = \hat{y}_a - \hat{y}_b + \hat{z}_b - \hat{z}_a$$
$$\geq 0 + \hat{z}_b - \hat{z}_a = \ell_g(x, y) + \ell_g(y, z).$$

Suppose that $\hat{x} \in R_b$, $\hat{z} \in R_b$, $\hat{y} \in H_b$, and $\hat{z}_a \geq \hat{z}_b$. Then

$$\ell_f(\hat{x}, \hat{y}) + \ell_f(\hat{y}, \hat{z}) = 0 + \hat{z}_b - \hat{z}_{ab} = 0 + \hat{z}_b - \hat{z}_b$$
$$= \ell_g(x, y) + \ell_g(y, z).$$

A similar argument applies for all the other cases.

The condition that Γ is finite means that Theorem 4.2.7 is restricted to deterministic allocation rules. If $|\Gamma|$ is not finite, additional conditions are needed.[10]

Example 5 *Here we give an example of a randomized allocation rule that satisfies the 2-cycle condition that is not IDS. Imagine two goods. The type (t_1, t_2) of an agent represents the value for each of the two goods. For convenience, suppose the value for not receiving any goods to be zero. Assume the type space to be $[0, 1]^2$. The allocation rule g specifies the probability of receiving each of the goods. Specifically, $g(t_1, t_2) = (\frac{t_1 + 2t_2}{3}, \frac{t_2}{3})$. If we let*

$$A = \begin{pmatrix} \frac{1}{3} & \frac{2}{3} \\ 0 & \frac{1}{3} \end{pmatrix}, \tag{4.11}$$

then $g(t) = At$ and $v(g(t)|t) = tAt$.

Observe that $\ell(s, t) = tA(t - s)$. To check that g satisfies the 2-cycle condition, it suffices to verify that

$$(s - t)A(s - t) \geq 0$$

for all types s and t. This task is left to the reader.

Consider now the following cycle:

$$(1, 0) \to (0, 1/2) \to (0, 1) \to (1, 0)$$

Now $\ell((0, 1), (1, 0)) = -\frac{1}{3}$, $\ell((1, 0), (0, 1/2)) = \frac{1}{12}$, and $\ell((0, 1/2), (0, 1)) = \frac{1}{6}$. Thus, the cycle has negative length, that is, g is not IDS.

Next we exhibit an extension of Theorem 4.2.7 to randomized allocation rules. A similar result appears in Archer and Kleinberg (2008) under more general conditions, and employs a different proof.

Suppose that Γ is finite and let $\Delta(\Gamma)$ denote the set of probability distributions over Γ. Notice that $\Delta(\Gamma)$ can be represented as a simplex in $\mathbb{R}^{|\Gamma|}$. If $g : T \mapsto \Delta(\Gamma)$ is an allocation rule, we can represent $g(t)$ for each t as a vector in the simplex. If the agent is risk-neutral, $v(g(t)|t) = t \cdot g(t)$, where $\cdot$ is the vector dot-product. Therefore $\ell(s, t) = t \cdot g(t) - t \cdot g(s) = t \cdot [g(t) - g(s)]$.

Lemma 4.2.9 *Suppose Γ is finite, $T \subseteq \mathbb{R}^{|\Gamma|}$ is convex, $v(\cdot|t)$ is a dot-product valuation, and $g : T \mapsto \Delta(\Gamma)$. Then, for any $s, t \in T$ and $\epsilon > 0$, there is a sequence $r^1, r^2, \ldots, r^k \in T$ such that*

$$\ell(s, r^1) + \ell(r^1, r^2) + \cdots + \ell(r^k, t) + \ell(t, r^k) + \cdots + \ell(r^1, s) \leq \epsilon.$$

Proof. Let $d = t - s$ and choose an integer k sufficiently large. Let $r^j = s + (\frac{j}{k+1})d$ and $\mu_j = g(r_j)$ for $j = 1, \ldots, k$. Set $r^0 = s$, $\mu_0 = g(s)$, $r^{k+1} = t$,

---

[10] In the classical approach, this is an integrability condition (see Rochet [1987]). We postpone discussion of this integrability condition to the section that covers the classical approach.

and $\mu_{k+1} = g(t)$. Observe that

$$\ell(r^j, r^{j+1}) + \ell(r^{j+1}, r^j) = (r^{j+1} - r^j) \cdot [g(r^{j+1}) - g(r^j)]$$
$$= \frac{d \cdot [g(r^{j+1}) - g(r^j)]}{k+1}.$$

Hence,

$$\ell(s, r^1) + \ell(r^1, r^2) + \cdots + \ell(r^k, t) + \ell(t, r^k) + \cdots + \ell(r^1, s)$$
$$= \frac{d \cdot [g(t) - g(s)]}{k+1} \leq \epsilon$$

for k sufficiently large. ∎

Theorem 4.2.10 *Suppose Γ is finite, $T \subseteq \mathbb{R}^{|\Gamma|}$ is convex, $v(\cdot|t)$ is a dot-product valuation, and $g : T \mapsto \Delta(\Gamma)$. If every finite cycle in T_g with at most three extreme points is nonnegative, then g is IDS.*

Proof. Suppose not. Then, by Theorem 4.2.1, T_g must contain a negative cycle

$$C : t^1 \to t^2 \to \cdots \to t^k \to t^1.$$

By the hypothesis of the theorem, C must have at least four extreme points. Pick two extreme points on C that are not adjacent on C. Without loss, call them t^1 and t^r. Now "split" the cycle C into two cycles:

$$C^1 : t^1 \to t^2 \to \cdots \to t^r \to [t^r, t^1]_\epsilon \to t^1$$

$$C^2 : t^r \to t^{r+1} \to \cdots \to t^1 \to [t^1, t^r]_\epsilon \to t^r$$

By Lemma 4.2.9,

$$\ell(C^1) + \ell(C^2) \leq \ell(C) + \epsilon < 0$$

for ϵ sufficiently small. Without loss, we may suppose that $\ell(C^1) < 0$. Notice that C^1 will have one less extreme point than C. ∎

As before, an equivalent version of the theorem is available.

Theorem 4.2.11 *Suppose Γ is finite, $T \subseteq \mathbb{R}^{|\Gamma|}$ is convex, $v(\cdot|t)$ a dot-product valuation, and $g : T \mapsto \Delta(\Gamma)$. If g is IDS on every two-dimensional subspace of T, then g is IDS.*

4.2.3 Convex Valuations

Here we consider a class of valuations that cannot be represented as a dot-product valuation, for which the previous results go through using essentially the same ideas.[11] We assume Γ is finite, $T \subseteq \mathbb{R}^{|\Gamma|}$ is convex, and $g : T \mapsto \Delta(\Gamma)$.

---

[11] This is based on Berger, Müller, and Naeemi (2009).

Because $v(\alpha|t)$ is assumed convex in $t \in T$, it cannot be expressed as a dot-product valuation.

The first step is to show that Lemma 4.2.9 holds. Basically, replacing linearity with convexity will mean replacing certain equalities by inequalities that go in the right direction. Recalling the notation of Lemma 4.2.9, we see that

$$\ell(r^j, r^{j+1}) = v(g(r^{j+1})|r^{j+1}) - v(g(r^j)|r^{j+1})$$

and

$$\ell(r^{j+1}, r^j) = v(g(r^j)|r^j) - v(g(r^{j+1})|r^j).$$

Therefore,

$$\ell(r^j, r^{j+1}) + \ell(r^{j+1}, r^j) = [v(g(r^{j+1})|r^{j+1}) - v(g(r^{j+1})|r^j)]$$
$$+ [v(g(r^j)|r^j) - v(g(r^j)|r^{j+1})].$$

Because $v(g(r^{j+1})|t)$ is convex in t, there exists a function $h_{j+1} : T \mapsto \mathbb{R}^{|\Gamma|}$, called the *subgradient* of v, such that

$$v(g(r^{j+1})|r^j) \geq v(g(r^{j+1})|r^{j+1}) + h_{j+1}(r^{j+1}) \cdot (r^j - r^{j+1}).$$

Rearranging the last inequality yields

$$v(g(r^{j+1})|r^{j+1}) - v(g(r^{j+1})|r^j) \leq h_{j+1}(r^{j+1}) \cdot (r^{j+1} - r^j).$$

Similarly,

$$v(g(r^j)|r^j) - v(g(r^j)|r^{j+1}) \leq -h_j(r^j) \cdot (r^{j+1} - r^j).$$

Hence,

$$\ell(r^j, r^{j+1}) + \ell(r^{j+1}, r^j) \leq \frac{d \cdot [h_{j+1}(r^{j+1}) - h_j(r^j)]}{k+1}.$$

Therefore,

$$\ell(s, r^1) + \cdots + \ell(r^k, t) + \ell(t, r^k) + \cdots + \ell(r^1, s)$$
$$= \frac{d \cdot [h_{k+1}(t) - h_0(s)]}{k+1} \leq \epsilon$$

for k sufficiently large.

To complete the argument, it suffices to note that Theorem 4.2.10 does not rely on a dot-product representation of the valuations. Hence, the following holds:

Theorem 4.2.12 *Suppose Γ is finite, $T \subseteq \mathbb{R}^{|\Gamma|}$ is convex, $v(\cdot|t)$ is convex in t, and $g : T \mapsto \Delta(\Gamma)$. If g is IDS on every two-dimensional subspace of T, then g is IDS.*

Implicit in the statement of Theorem 4.2.12 is that T is at least two-dimensional. What if T is one-dimensional? Convexity of valuations combined with the 2-cycle condition is not sufficient to ensure that an allocation rule g is IDS in one dimension. The example below illustrates this.

4.2 Dominant Strategy Incentive Compatibility

Example 6 *Suppose $T = [0, 1]$. Consider the following allocation function:*

$$f(t) = \begin{cases} a & 0 \leq t \leq \frac{1}{3} \\ b & \frac{1}{3} < t \leq \frac{2}{3} \\ c & \frac{2}{3} < t \leq 1. \end{cases}$$

We define the corresponding valuation functions:

$$v(a, t) = \begin{cases} 0 & t \leq \frac{2}{3} \\ 3t - 2 & t > \frac{2}{3} \end{cases}$$

$v(b, t) = 3t$ *and*

$$v(c, t) = \begin{cases} 2 - 3t & t \leq \frac{1}{3} \\ 3t & t > \frac{1}{3} \end{cases}$$

It is easy to check that the function f satisfies the 2-cycle condition but is not IDS.

Contrast this with Theorem 4.2.5 where strict increasing differences implied that the 2-cycle condition was sufficient for IDS.

The example reveals that convexity of the valuations is not sufficient to bound the length of the shortest cycle using information only about the 2-cycles. For the argument to go through, the allocation rule g must possess another property defined below.

An allocation rule is **decomposition monotone** if $\forall \underline{r}, \bar{r} \in T$ and $\forall r \in T$ s.t. $r = (1 - \alpha)\underline{r} + \alpha \bar{r}, \alpha \in (0, 1)$ we have

$$\ell\left(\underline{r}, \bar{r}\right) \geq \ell\left(\underline{r}, r\right) + \ell(r, \bar{r}). \tag{4.12}$$

If decomposition monotonicity holds, the arc between those nodes is at least as long as any path connecting the same two nodes via nodes lying on the line segment between them. Figure 4.2 gives an illustrative example. Decomposition monotonicity implies that the arc from $\underline{r}$ to $\bar{r}$ is at least as long as the path $A = \left(\underline{r}, r_{**}, \bar{r}\right)$, and that A is at least as long as the path $\tilde{A} = \left(\underline{r}, r_*, r_{**}, r_{***}, \bar{r}\right)$. Valuations that can be represented as dot-products are clearly decomposition monotone. The allocation rule of Example 6 fails to satisfy decomposition monotonicity.

With decomposition monotonicity, the following is straightforward to prove.

Theorem 4.2.13 *Suppose Γ is finite, $T \subseteq \mathbb{R}^{|\Gamma|}$ is convex, $v(\cdot|t)$ convex in t, and $g : T \mapsto \Delta(\Gamma)$ is decomposition monotone. If g is IDS on every one-dimensional subspace of T, then g is IDS.*

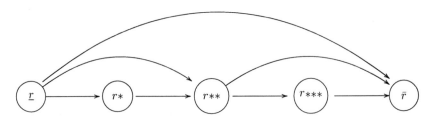

Figure 4.2 Decomposition monotonicity.

4.2.4 Roberts's Theorem

In this section, we provide proof of a theorem of Roberts (1979) that gives a characterization of IDS allocation rules as solutions to maximization problems. The proof is a reinterpretation of the original emphasizing the underlying network structure.[12] It lays bare the precise points where a crucial assumption about the type space is used and allows a slight strengthening.

Theorem 4.2.14 (Roberts's Theorem) *Suppose Γ is finite and contains at least three elements.[13] Let $T = \mathbb{R}_+^{|\Gamma|}$ and g be IDS and onto. Then there exists a nonzero, nonnegative $w \in \mathbb{R}^n$ and $|\Gamma|$ real numbers $\{D_\gamma\}_{\gamma \in \Gamma}$ such that*

$$g(\mathbf{t}) \in \arg\max_{\gamma \in \Gamma} \sum_{i=1}^{n} w_i t_\gamma^i - D_\gamma.$$

The class of allocation rules described in Roberts's Theorem are called **affine maximizers**. Another way to state Roberts's Theorem is that there is a solution $w, \{D_\gamma\}_{\gamma \in \Gamma}$ to the following system:

$$D_\alpha - D_\gamma \leq \sum_{i=1}^{n} w_i(t_\alpha^i - t_\gamma^i) \ \forall \gamma, \ \mathbf{t} \text{ s.t. } g(\mathbf{t}) = \alpha.$$

This immediately suggests a network representation. Fix a nonzero and nonnegative vector w. We associate a network with w called Γ_w. The network will have one node for each $\gamma \in \Gamma$. For each ordered pair (β, α) where $\beta, \alpha \in \Gamma$, introduce a directed arc from β to α of length

$$l_w(\beta, \alpha) = \inf_{\mathbf{t}: g(\mathbf{t}) = \alpha} \sum_{i=1}^{n} w_i(t_\alpha^i - t_\beta^i).$$

If there is a choice of w for which Γ_w has no negative length cycles, Roberts's Theorem is proven. We will prove the following:

Theorem 4.2.15 *Let g be IDS, onto and $|\Gamma| \geq 3$. Then there exists a nonzero, nonnegative $w \in \mathbb{R}^n$ such that Γ_w has no negative length cycles.*

---

[12] See also Lavi, Mu'alem, and Nisan (2004) and Meyer-ter-Vehn and Moldovanu (2002).
[13] It is not hard to extend the theorem to the case when Γ is infinite.

4.2 Dominant Strategy Incentive Compatibility

In what follows, we can assume wlog that $w \geq 0$ and $\sum w_i = 1$. Such a vector w will be called *feasible*.

First we give an outline of the proof. Suppose a cycle $C = \alpha_1 \to \cdots \to \alpha_k \to \alpha_1$ through the elements of Γ. From each α_j, pick a profile $\mathbf{t}[j]$ such that $g(\mathbf{t}[j]) = \alpha_j$. We associate with the cycle C a vector b whose i^{th} component is

$$b^i = (t^i_{\alpha_1}[1] - t^i_{\alpha_k}[1]) + (t^i_{\alpha_2}[2] - t^i_{\alpha_1}[2]) + \cdots + (t^i_{\alpha_k}[k] - t^i_{\alpha_{k-1}}[k]).$$

Let $K \subseteq \mathbb{R}^n$ be the set of vectors that can be associated with some cycle through the elements of Γ. Theorem 4.2.15 asserts the existence of a feasible w such that $w \cdot b \geq 0$ for all $b \in K$. The major milestones of the proof are as follows:

(1) If $b \in K$ is associated with the cycle $\alpha_1 \to \cdots \to \alpha_k \to \alpha_1$, then b is associated with the cycle $\alpha_1 \to \alpha_k \to \alpha_1$. In words, each element of K is associated with a cycle through a pair of elements of Γ.
(2) If $b \in K$ is associated with a cycle through (α, β), then b is associated with a cycle through (γ, θ) for all $(\gamma, \theta) \neq (\alpha, \beta)$. In words, attention can be restricted to just one cycle.
(3) The set K is convex.
(4) Finally, it suffices to note that K is disjoint from the negative orthant. Theorem 4.2.15 follows by invoking the separating hyperplane theorem.

Let $U(\beta, \alpha) = \{d \in \mathbb{R}^n : \exists \mathbf{t} \in T^n \text{ s.t. } g(\mathbf{t}) = \alpha, \text{ s.t. } d^i = t^i_\alpha - t^i_\beta \ \forall i\}$. Because g is onto, $U(\beta, \alpha) \neq \emptyset$ for all $\alpha, \beta \in \Gamma$. Notice that $l_w(\beta, \alpha) = \inf_{d \in U(\beta,\alpha)} w \cdot d$.

Lemma 4.2.16 *Suppose $g(\mathbf{t}) = \alpha$ and $\mathbf{s} \in T^n$ such that $s^i_\alpha - s^i_\beta > t^i_\alpha - t^i_\beta$ for all i. Then $g(\mathbf{s}) \neq \beta$.*

Proof. This lemma is a consequence of the 2-cycle inequality, and we leave its proof as an exercise. As a hint, consider the profile $(s^1, \mathbf{t}^{-1})$ and suppose that $s^1_\alpha - s^1_\beta > t^1_\alpha - t^1_\beta$ and $g(s^1, \mathbf{t}^{-1}) = \beta$. ∎

A consequence of Lemma 4.2.16 is that $U(\beta, \alpha)$ is upper comprehensive for all $\alpha, \beta \in \Gamma$.[14] To see why, consider any profile $\mathbf{s}$ such that $g(\mathbf{s}) = \alpha$ and $d^i = s^i_\alpha - s^i_\beta$ for all i. Choose a $\hat{d} \geq d$. Consider a profile $\mathbf{t}$ such that

$$t^i_\alpha - t^i_\beta = \hat{d}^i \ \forall i$$

and

$$t^i_\alpha - t^i_\gamma > s^i_\alpha - s^i_\gamma \ \forall i, \ \forall \gamma \neq \alpha, \beta.$$

It is easy to verify that such a profile exists. Repeated application of Lemma 4.2.16 implies that $g(\mathbf{t}) = \alpha$.

---

[14] A set $Y \subset \mathbb{R}^k$ is upper comprehensive if whenever $y \in Y$, and $z \geq y$, then $z \in Y$.

For every pair $\alpha, \beta \in \Gamma$ define
$$h(\beta, \alpha) = \inf_{t \in T^n : g(t) = \alpha} \max_i t_\alpha^i - t_\beta^i = \inf_{d \in U(\beta, \alpha)} \max_i d^i.$$

Lemma 4.2.17 *For every pair $\alpha, \beta \in \Gamma$, $h(\beta, \alpha)$ is finite.*

Proof. Suppose not. Fix a pair α and β for which the lemma is false. Then $h(\beta, \alpha)$ can be made arbitrarily small. Because $U(\beta, \alpha)$ is upper comprehensive, this would imply that $U(\beta, \alpha) = \mathbb{R}^n$.

Choose $d \in U(\alpha, \beta)$. Then there is an $s \in T^n$ such that $s_\beta^i - s_\alpha^i = d^i$ for all i and $g(s) = \beta$. Because $U(\beta, \alpha) = \mathbb{R}^n$, there is a $t \in T^n$ such that $t_\alpha^i - t_\beta^i < -d^i$ for all i and $g(t) = \alpha$. Because $t_\beta^i - t_\alpha^i > s_\beta^i - s_\alpha^i$ for all i, it follows from Lemma 4.2.16 that $g(t) \neq \alpha$, which is a contradiction. ∎

Observe that if for some $t \in T^n$ we have that $t_\alpha^i - t_\beta^i < h(\beta, \alpha)$ for all i, then $g(t) \neq \alpha$.

Lemma 4.2.18 *For all $\alpha, \beta \in \Gamma$, $h(\alpha, \beta) + h(\beta, \alpha) = 0$.*

Proof. Suppose first that $h(\alpha, \beta) + h(\beta, \alpha) > 0$. Choose $t \in T^n$ to satisfy

$$t_\alpha^i - t_\beta^i < h(\beta, \alpha) \ \forall i \tag{4.11}$$

$$t_\beta^i - t_\alpha^i < h(\alpha, \beta) \ \forall i \tag{4.12}$$

$$t_\gamma^i - t_\alpha^i < h(\alpha, \gamma) \ \forall i \ \forall \gamma \neq \alpha, \beta. \tag{4.13}$$

It is is easy to see that the network associated with this system has no negative length cycle. So, it has a solution. Further, adding the same constant to each t_γ^i for all i and $\gamma \in \Gamma$ does not violate feasibility. Hence, we may assume the solution is nonnegative.

Now (4.11) implies that $g(t) \neq \alpha$. Similarly, (4.12) implies that $g(t) \neq \beta$. Together with (4.13), we deduce that $g(t) \notin \Gamma$ is a contradiction.

Now suppose that $h(\alpha, \beta) + h(\beta, \alpha) < 0$. Choose $t \in T^n$ to satisfy

$$t_\beta^i - t_\alpha^i > h(\alpha, \beta) \ \forall i \tag{4.14}$$

$$t_\alpha^i - t_\beta^i > h(\beta, \alpha) \ \forall i \tag{4.15}$$

$$t_\gamma^i - t_\alpha^i < h(\alpha, \gamma) \ \forall i \ \forall \gamma \neq \alpha, \beta. \tag{4.16}$$

It is easy to see that the system has a nonnegative solution. Now (4.16) implies that $g(t) \in \{\alpha, \beta\}$. However, (4.15) coupled with Lemma 4.2.16 means that $g(t) \neq \beta$. But (4.14) and lemma 4.2.16 imply that $g(t) \neq \alpha$ is a contradiction. ∎

Denote the vector all of whose components equal 1 by $\hat{u}$.

Lemma 4.2.19 *If $d_{\gamma, \beta} \in U(\gamma, \beta)$, $d_{\beta, \alpha} \in U(\beta, \alpha)$, then for all $\epsilon > 0$,*

$$d_{\gamma, \beta} + d_{\beta, \alpha} + \epsilon \hat{u} \in U(\gamma, \alpha).$$

4.2 Dominant Strategy Incentive Compatibility

Proof. Choose $\mathbf{t} \in T^n$ to satisfy the following:

$$t^i_\alpha - t^i_\beta = d^i_{\beta,\alpha} + \epsilon/2 \quad \forall i \tag{4.17}$$

$$t^i_\beta - t^i_\gamma = d^i_{\gamma,\beta} + \epsilon/2 \quad \forall i \tag{4.18}$$

$$t^i_\theta - t^i_\alpha < h(\alpha, \theta) \quad \forall \theta \notin \{\alpha, \beta, \gamma\}, \; \forall i. \tag{4.19}$$

Because the network associated with the system has no cycle, the system has a solution. In fact, by adding constants, we can ensure a nonnegative solution. Inequality (4.19) combined with the definition of h implies that $g(\mathbf{t}) \in \{\alpha, \beta, \gamma\}$. Inequality (4.18) and Lemma 4.2.16 imply that $g(\mathbf{t}) \neq \gamma$. Inequality (4.17) combined with Lemma 4.2.16 imply that $g(\mathbf{t}) \neq \beta$. Hence, $g(\mathbf{t}) = \alpha$. Observe that $t^i_\alpha - t^i_\gamma = d^i_{\beta,\alpha} + d^i_{\gamma,\beta} + \hat{u}\epsilon$ for all i, and this proves the lemma.

If the length function l_w is well defined for all arcs of Γ_w, the conclusion of Lemma 4.2.19 can be restated as

$$l_w(\alpha, \beta) \leq l_w(\alpha, \gamma) + l_w(\gamma, \beta) + \epsilon.$$

In words, l_w satisfies the triangle inequality.

Lemma 4.2.20 *Suppose for some feasible w we have that $l_w(\alpha, \beta) + l_w(\beta, \gamma) \geq 0$ for all $\alpha, \beta \in \Gamma$. Then Γ_w has no negative length cycles.*

Proof. The condition that $l_w(\alpha, \beta) + l_w(\beta, \gamma) \geq 0$ for all $\alpha, \beta \in \Gamma$ implies that the length function l_w is finite. To prove the lemma, it suffices to show that l_w satisfies the triangle inequality. This follows from Lemma 4.2.19. ∎

Lemma 4.2.21 *For any $d_{\gamma,\alpha} \in U(\gamma, \alpha)$, $d_{\alpha,\beta} \in U(\alpha, \beta)$, and $d_{\beta,\alpha} \in U(\beta, \alpha)$, we have*

$$d_{\gamma,\alpha} + d_{\alpha,\beta} + d_{\beta,\alpha} + 2\hat{u}\epsilon \in U(\gamma, \alpha).$$

Proof. Lemma 4.2.19 implies that $d_{\gamma,\alpha} + d_{\alpha,\beta} + \hat{u}\epsilon \in U(\gamma, \beta)$. Applying Lemma 4.2.19 again yields

$$d_{\gamma,\alpha} + d_{\alpha,\beta} + \hat{u}\epsilon + d_{\beta,\alpha} + \hat{u}\epsilon \in U(\gamma, \alpha).$$

∎

Because $U(\alpha, \beta) \neq \mathbb{R}^n$, it follows that the complement of $U(\alpha, \beta)$ must contain all vectors x such that $-x + \epsilon \hat{u} \in U(\beta, \alpha)$. In particular, this means that $int[U(\alpha, \beta) + U(\beta, \alpha)]$ does not contain the origin.

Lemma 4.2.22 $int[U(\alpha, \beta) + U(\beta, \alpha)]$ *is convex for all $\alpha, \beta \in \Gamma$.*

Proof. Pick $d_{\alpha,\beta}, d'_{\alpha,\beta} \in U(\alpha, \beta)$ and $d_{\beta,\alpha}, d'_{\beta,\alpha} \in U(\beta, \alpha)$. It suffices to prove that

$$(d_{\alpha,\beta} + d_{\beta,\alpha})/2 + (d'_{\alpha,\beta} + d'_{\beta,\alpha})/2 \in int[U(\alpha, \beta) + U(\beta, \alpha)].$$

Suppose not. Then, without loss, we may suppose that $(d_{\alpha,\beta} + d'_{\alpha,\beta})/2 \notin int U(\alpha, \beta)$. Hence, $-(d_{\alpha,\beta} + d'_{\alpha,\beta})/2 \in U(\beta, \alpha)$. Therefore, by Lemma 4.2.21,

$$h(\gamma, \alpha)\hat{u} + \epsilon \hat{u} + d_{\alpha,\beta} - (d_{\alpha,\beta} + d'_{\alpha,\beta})/2$$
$$= h(\gamma, \alpha)\hat{u} + \hat{u}\epsilon + (d_{\alpha,\beta} - d'_{\alpha,\beta})/2 \in U(\gamma, \alpha).$$

Because $h(\alpha, \gamma)\hat{u} + \hat{u}\epsilon \in U(\alpha, \gamma)$, it follows from Lemma 4.2.18

$$(d_{\alpha,\beta} - d'_{\alpha,\beta})/2 + 2\hat{u}\epsilon = (d_{\alpha,\beta} - d'_{\alpha,\beta})/2 + \hat{u}\epsilon$$
$$+ h(\gamma, \alpha)\hat{u} + h(\alpha, \gamma)\hat{u} + \hat{u}\epsilon \in int[U(\alpha, \gamma) + U(\gamma, \alpha)].$$

By Lemma 4.2.21,

$$-d_{\alpha,\beta}\hat{u}\epsilon + (d_{\alpha,\beta} - d'_{\alpha,\beta})/2 + 2\hat{u}\epsilon \in U(\alpha, \beta)$$
$$\Rightarrow -(d_{\alpha,\beta} + d'_{\beta,\alpha})/2 \in int[U(\alpha, \beta)],$$

which contradicts the hypothesis that $-(d_{\alpha,\beta} + d'_{\alpha,\beta})/2 \in U(\beta, \alpha)$. ∎

Lemma 4.2.23 *If there is a feasible w and pair $\alpha, \beta \in \Gamma$ such that $l_w(\alpha, \beta) + l_w(\beta, \alpha) < 0$, then for all $\gamma, \theta \in \Gamma$, we have $l_w(\gamma, \theta) + l_w(\theta, \gamma) < 0$.*

Proof. It suffices to prove that $int[U(\alpha, \beta) + U(\beta, \alpha)] = int[U(\gamma, \delta) + U(\delta, \gamma)]$ for all $\alpha, \beta, \gamma, \delta \in \Gamma$. For any $d_{\alpha,\beta} \in U(\alpha, \beta)$, we have by Lemma 4.2.19 that $d_{\alpha,\beta} + \hat{u}h(\beta, \gamma) + \hat{u}\epsilon \in U(\alpha, \gamma)$. Similarly, for any $d_{\beta,\alpha} \in U(\beta, \alpha)$, we have that $d_{\beta,\alpha} + h(\gamma, \beta)\hat{u} + \hat{u}\epsilon \in U(\gamma, \alpha)$. Hence, by Lemma 4.2.18,

$$d_{\alpha,\beta} + h(\beta, \gamma)\hat{u} + \hat{u}\epsilon + d_{\beta,\alpha} + h(\gamma, \beta)\hat{u} + \hat{u}\epsilon$$
$$= (d_{\alpha,\beta} + \hat{u}\epsilon) + (d_{\beta,\alpha} + \hat{u}\epsilon) \in U(\alpha, \gamma) + U(\gamma, \alpha).$$

Hence, $int[U(\alpha, \beta) + U(\beta, \alpha)] \subseteq U(\alpha, \gamma) + U(\gamma, \alpha)$. Switching the roles of β and α, we conclude that $int[U(\alpha, \beta) + U(\beta, \alpha)] = int[U(\alpha, \gamma) + U(\gamma, \alpha)]$. Repeating the argument again with α replaced by δ completes the proof. ∎

We now prove Theorem 4.2.15. Suppose it is false. Then, by Lemma 4.2.23, for every feasible choice of w and pair $\alpha, \beta \in \Gamma$, we have $l_w(\alpha, \beta) + l_w(\beta, \alpha) < 0$. By Lemma 4.2.22, the set $int(U(\alpha, \beta)) + int(U(\beta, \alpha))$ is convex. Notice also that the set $int(U(\alpha, \beta)) + int(U(\beta, \alpha))$ cannot contain the origin. Therefore, by the separating hyperplane theorem, there is a feasible w^* such that $w^* \cdot z \geq 0$ for all $z \in U(\alpha, \beta) + U(\beta, \alpha)$, that is a contradiction.

Theorem 4.2.15 is false when $|\Gamma| = 2$ and when T is a strict subset of $\mathbb{R}^{|\Gamma|}$.

Example 7 We describe a counterexample to Theorem 4.2.15 when $|\Gamma| = 2$. Suppose $\Gamma = \{\alpha, \beta\}$. Let $g(\mathbf{t}) = \alpha$ whenever $t^i_\alpha \geq t^i_\beta$ for all agents i, otherwise $g(\mathbf{t}) = \beta$. Clearly, g cannot be expressed as an affine maximizer. To show that g is IDS, we verify that the 2-cycle condition holds. Fix $\mathbf{t}^{-i}$. We have two cases. In the first, $t^j_\alpha \geq t^j_\beta$ for all $j \neq i$. Then the corresponding graph Γ_g consists of just two nodes; α and β. As long as $t^i_\alpha - t^i_\beta \geq 0$, we know that $g(t^i, \mathbf{t}^{-i}) = \alpha$. Hence, $\ell(\beta, \alpha) = 0$. Similarly, $\ell(\alpha, \beta) = 0$. In the second case, $t^j_\alpha < t^j_\beta$ for at

4.3 Revenue Equivalence

least one $j \neq i$. In this case, $g(t^i, \mathbf{t}^{-i}) = \beta$ for all t^i. Hence, the corresponding graph Γ_g is a single vertex.

Example 8 *We give a counterexample to Theorem 4.2.15 when T is a strict subset of $\mathbb{R}^{|\Gamma|}$. Consider two buyers and one good. There are three possible allocations: the good goes to agent 1, it goes to agent 2, or it is not allocated at all. An agents type records the value of each possible allocation. Suppose private values, that is, agent i receives value of $v_i \in [0, 1]$ when she receives the good and zero for all other allocations. Hence, T is a one-dimensional subset of $[0, 1]^3$. Consider the following allocation rule. If $v_1, v_2 \in [0, 1/2)$, the good is not allocated. When $v_1, v_2 \in (1/2, 1]$, the good is not allocated. In all other cases, the good is allocated to the agent with the highest value. It is straightforward to verify that the rule is IDS and not an affine maximizer.*

The assumption that $T = \mathbb{R}_+^{|\Gamma|}$ is used in two substantial ways in the proof of Theorem 4.2.15: Lemma 4.2.18 and Lemma 4.2.19.[15] In each case, a solution to a system of inequalities (that are dual to a shortest-path problem) needed to be checked for feasibility. Under the assumption that $T = \mathbb{R}_+^{|\Gamma|}$, it was enough to verify that the associated network contained no negative cycle. However, if additional restrictions were imposed on T, one would have to check that among the solutions whose existence is verified, there is at least one that satisfied the additional restrictions. When those additional restrictions can be formulated using inequalities that are dual to a shortest-path problem, this is easy to do. We outline one example.

Suppose a public good setting where the mechanism designer must choose a subset S from a ground set M. This set may represent some combination of infrastructure. The utility that agent i derives from the choice S is $u_i(S)$. In this setting, Γ is the set of all subsets of M. The type of an agent i is $\{u_i(S)\}_{S \subseteq M}$. In the absence of restrictions on the $u_i(\cdot)$'s, the type space is all of $\mathbb{R}_+^{|\Gamma|}$, and so Theorem 4.2.15 applies. Suppose we require that the $u_i(\cdot)$ be monotone for all i. Now, the type space will be a lower-dimensional subset of $\mathbb{R}_+^{|\Gamma|}$, and Theorem 4.2.15 does not obviously hold. Nevertheless, it does. To prove this, it suffices to show that the types chosen in the proofs of Lemma 4.2.18 and Lemma 4.2.19 can be selected so as to satisfy monotonicity constraints. These monotonicity constraints take the form $t_\alpha - t_\beta \leq 0$ if $\alpha \subset \beta$. Observe that these inequalities have the appropriate network structure. We leave it to the reader to verify the details.

4.3 REVENUE EQUIVALENCE

Suppose T is a type space and g an allocation rule that is IDS. **Revenue equivalence** is said to hold for g on T if for any pair of payment schemes P and $\tilde{P}$ such that (g, P) and $(g, \tilde{P})$ are IDS mechanisms and for each agent i and

---

[15] The original version of Theorem 4.2.15 assumes $T = \mathbb{R}^{|\Gamma|}$.

report $\mathbf{t}^{-i}$ of the other agents, there exists $h_i(\mathbf{t}^{-i}) \in \mathbb{R}$ such that $\tilde{P}_i(t^i, \mathbf{t}^{-i}) = P_i(t^i, \mathbf{t}^{-i}) + h_i(\mathbf{t}^{-i})$. In other words, any two payment rules differ by a constant.

Now let us revert again to the more compact notation where the dependence on $\mathbf{t}^{-i}$ is suppressed. If (g, P) is dominant strategy incentive compatible, we have, by the Taxation Principle, that for any pair of types $t, s \in T$ such that $g(t) = g(s) = \gamma$ for some $\gamma \in \Gamma$, $P(t) = P(s)$. In particular, payments can be indexed by allocation rather than type. That is, for all t such that $g(t) = \alpha$, $P(t) = P^\alpha$. In this notation, revenue equivalence can be stated this way. Choose a $\alpha \in \Gamma$ and fix the value of P^α. Then the values of P^γ for all $\gamma \in \Gamma \setminus \alpha$ are uniquely determined.

The bulk of prior work on revenue equivalence identifies sufficient conditions on the type space for *all* allocation rules from a certain class to satisfy revenue equivalence. Green and Laffont (1977) and Holmstrom (1979) restrict attention to allocation rules called utilitarian maximizers, that is, allocation rules that maximize the sum of the valuations of all agents. They show that when the type space is smoothly path-connected, then utilitarian maximizers satisfy revenue equivalence.

Myerson (1981) shows that revenue equivalence holds for every implementable rule in a setting where the type space is an interval of the real line, the outcome space is a lattice, and an agent's valuation for an outcome is continuous and supermodular in her type.

Krishna and Maenner (2001) derive revenue equivalence under two different hypotheses. In the first, agents' type spaces must be convex and the valuation function of an agent is a convex function of the type of the agent. Under these conditions, they show that every implementable rule satisfies revenue equivalence. The second hypothesis requires the allocation rule to satisfy certain differentiability and continuity conditions and the outcome space to be a subset of the Euclidean space. Furthermore, the valuation functions must be regular Lipschitzian and monotonically increasing in all arguments.[16]

Milgrom and Segal (2002) show that revenue equivalence is a consequence of a particular envelope theorem in a setting where the type spaces are one-dimensional and the outcome space is arbitrary. An agent's valuation function is assumed differentiable and absolutely continuous in the type of the agent, and the partial derivative of the valuation function with respect to the type must satisfy a certain integrability condition. Their result can be applied to multidimensional type spaces as well. In this case, the type spaces must be smoothly connected and the valuation functions must be differentiable with bounded gradient.

There are two papers that identify necessary as well as sufficient conditions for revenue equivalence to hold. If the outcome space is finite, Suijs (1996) characterizes type spaces and valuation functions for which utilitarian maximizers satisfy revenue equivalence. Chung and Olszewski (2007) characterize

---

[16] A function is regular at a point x if it admits one-sided directional derivatives in all directions at x.

4.3 Revenue Equivalence

type spaces and valuation functions for which *every* implementable allocation rule satisfies revenue equivalence, again under the assumption of a finite outcome space. From their characterization, they derive sufficient conditions on the type spaces and valuation functions that generalize known results when the outcome space is countable or a probability distribution over a finite set of outcomes.

Here we describe conditions on both the allocation rule g and the type space T that characterize when revenue equivalence holds.[17] The characterization presented here differs from prior work in that a joint condition on the type spaces – the valuation functions *and* the implementable allocation rule – is given that characterizes revenue equivalence. This characterization differs from the one in Chung and Olszewski (2007) in three ways. First, it holds for general outcome spaces. Second, it implies revenue equivalence in cases where their result does not apply. In fact, given agents' type spaces and valuation functions, several allocation rules may be implementable in dominant strategies, some of which satisfy revenue equivalence and some do not. In this case, the conditions on the type space and valuation functions from their paper obviously cannot hold. However, the present characterization can be used to determine which of the allocation rules do satisfy revenue equivalence. Third, the characterization in Chung and Olszewski (2007) is a corollary of the present one in the sense that their necessary and sufficient condition is naturally related to the graph theoretic interpretation of revenue equivalence.

Theorem 4.3.1 *Let $T \subseteq \mathbb{R}^{|\Gamma|}$ and g be IDS. For every pair $\alpha, \beta \in \Gamma$ denote by $L(\alpha, \beta)$ the shortest-path length from α to β in Γ_g. Revenue equivalence holds for g on T iff. $L(\alpha, \beta) + L(\beta, \alpha) = 0$ for all $\alpha, \beta \in \Gamma$.*

Proof. Since g is IDS, the network Γ_g has no negative length cycles. Therefore, shortest-path lengths in Γ_g are well defined. Choose a $\theta \in \Gamma$ and set P_θ^γ equal to the length of the shortest path from θ to $\gamma \in \Gamma$.[18] Notice that (g, P_θ) is dominant strategy incentive compatible for all choices of $\theta \in \Gamma$. Suppose revenue equivalence holds. Then for each $\mu, \nu \in \Gamma$, there is a constant c such that $P_\mu^\gamma - P_\nu^\gamma = c$ for all $\gamma \in \Gamma$. In particular,

$$P_\alpha^\alpha - P_\beta^\alpha = P_\alpha^\beta - P_\beta^\beta. \tag{4.20}$$

Using the fact that $P_\alpha^\alpha = P_\beta^\beta = 0$, equation (4.20) reduces to

$$-P_\beta^\alpha = P_\alpha^\beta \Rightarrow 0 = P_\alpha^\beta + P_\beta^\alpha = L(\alpha, \beta) + L(\beta, \alpha). \tag{4.21}$$

Now suppose that $L(\alpha, \beta) + L(\beta, \alpha) = 0$ for all $\alpha, \beta \in \Gamma$, and let P and $\hat{P}$ be payment rules such that (g, P) and $(g, \hat{P})$ are IDS. Observe first that for any

---

[17] It is based on Heydenreich, Müller, Uetz, and Vohra (2008).
[18] More precisely, the infimum over all finite path lengths. Since g is IDS, this is well defined. See the proof of Theorem 4.2.1.

finite path $\alpha_1 \to \alpha_2 \to \alpha_3 \to \cdots \to \alpha_k$, we must have by IDS that

$$P^{\alpha_k} - P^{\alpha_1} \le \ell(\alpha_1, \alpha_2) + \cdots + \ell(\alpha_{k-1}, \alpha_k).$$

This follows from the fact that $P^{\alpha_j} - P^{\alpha_{j-1}} \le \ell(\alpha_{j-1}, \alpha_j)$ for $j = 2, \ldots, k$. The same holds for $\hat{P}$. Invoking this inequality for all finite paths between α and β, we conclude that

$$P^\alpha - P^\beta \le L\beta, \alpha)$$

and

$$\hat{P}^\beta - \hat{P}^\alpha \le L(\alpha, \beta).$$

Adding these together yields: $\hat{P}^\beta - P^\beta \le \hat{P}^\alpha - P^\alpha$. Reversing the roles of P and $\hat{P}$ returns the opposite inequality, that is, $\hat{P}^\beta - P^\beta = \hat{P}^\alpha - P^\alpha$. ∎

An immediate implication of Theorem 4.3.1 and the discussion in section 4.2.3 is that revenue equivalence holds in the following case: Γ is finite, $T \subseteq \mathbb{R}^{|\Gamma|}$ is convex, $g : T \mapsto \Delta(\Gamma)$ and $v(\alpha|t)$ is convex in $t \in T$.

Theorem 4.3.1 imposes conditions on the pair (g, T) to conclude that revenue equivalence holds for g. Chung and Olszewski (2007), on the other hand, characterize the type space for which *every* IDS allocation rule satisfies revenue equivalence. It is an easy consequence of Theorem 4.3.1.

To describe the condition on type spaces introduced in Chung and Olszewski (2007), some notation will be required. Let B_1, B_2 be disjoint subsets of Γ and $r : B_1 \cup B_2 \to \mathbb{R}$. For every $\epsilon > 0$, let

$$V_1(\epsilon) = \cup_{b \in B_1} \{t \in T : \forall_{a \in B_2} : t_b - t_a > r(b) - r(a) + \epsilon\}$$

and

$$V_2(\epsilon) = \cup_{a \in B_2} \{t \in T : \forall_{b \in B_1} : t_b - t_a < r(b) - r(a) - \epsilon\}.$$

Finally, $V_i = \cup_{\epsilon > 0} V_i(\epsilon)$. Observe that $V_1 \cap V_2 = \emptyset$. Call a set T **splittable** if there are B_1, B_2 and r such that $T = V_1 \cup V_2$ where $T \cap V_i \ne \emptyset$ for $i = 1, 2$.

Theorem 4.3.2 *Let g be IDS and Γ finite. Then the following are equivalent:*

(1) g satisfies revenue equivalence.
(2) T is not splittable.

In Theorem 4.3.1, no assumption on the cardinality of Γ is made, whereas in Theorem 4.3.2, Γ is assumed finite. When Γ is not finite but countably infinite, it is shown in Chung and Olszewski (2007) that item 2 of Theorem 4.3.2 implies revenue equivalence. To prove that Statement 2 in Theorem 4.3.2 is a necessary condition for revenue equivalence in the case of finite Γ, one can directly construct an allocation rule and two payment schemes that do not differ by a constant from the assumption that T is splittable. Here we confine ourselves to an alternative proof of the fact that Statement 2 is a sufficient condition when Γ is countable. This makes clear the connection to the network Γ_g and assigns an interpretation to the function r defined above.

4.3 Revenue Equivalence

Let g be an allocation rule implementable in dominant strategies that violates revenue equivalence. Because g is IDS, the network Γ_g satisfies the nonnegative cycle property. Theorem 4.3.1 implies that there exist $\alpha, \beta \in \Gamma$ such that $L(\alpha, \beta) + L(\beta, \alpha) > 0$. We show that this implies that T is splittable.

Fix any $\alpha \in \Gamma$ and define $d(\beta) = L(\alpha, \beta) + L(\beta, \alpha)$ for all $\beta \in \Gamma$. Because the function d takes only countably many values, there exists $z \in \mathbb{R}$ such that the following sets form a nontrivial partition of Γ: $B_1 = \{\beta \in \Gamma \mid d(\beta) > z\}$, $B_2 = \{\beta \in \Gamma \mid d(\beta) < z\}$. Observe that for every $\gamma_1 \in B_1$, there exists $\epsilon(\gamma_1) > 0$ such that $d(\gamma_1) > z + \epsilon(\gamma_1)$. Similarly, for every $\gamma_2 \in B_2$, there exists $\epsilon(\gamma_2) > 0$ such that $d(\gamma_2) < z - \epsilon(\gamma_2)$.

For $\gamma_1 \in B_1$, let $r(\gamma_1) = -\ell(\pi(\gamma_1, \alpha))$. For $\gamma_2 \in B_2$, let $r(\gamma_2) = L(\alpha, \gamma_2) - z$. Now let $s \in T$ such that $g(s) = \gamma_1 \in B_1$. We claim that for all $\gamma_2 \in B_2$, it holds $v(\gamma_1|s) - v(\gamma_2|s) > r(\gamma_1) - r(\gamma_2) + \epsilon(\gamma_1)$, which proves $s \in V_1(\epsilon(\gamma_1))$. Indeed, $v(\gamma_1|s) - v(\gamma_2|s) \geq \ell(\gamma_2, \gamma_1)$. The claim then follows from:

$$\ell(\gamma_2, \gamma_1) + L(\gamma_1, \alpha) + L(\alpha, \gamma_2) \geq L(\alpha, \gamma_1) + L(\gamma_1, \alpha)$$
$$> L(\alpha, \gamma_1) - L(\alpha, \gamma_1)$$
$$\quad + z + \epsilon(\gamma_1)$$
$$= z + \epsilon(\gamma_1).$$

Next, let $s \in T$ such that $g(s) = \gamma_2 \in B_2$. We claim that for all $\gamma_1 \in B_1$, it holds $v(\gamma_1|s) - v(\gamma_2|s) < r(\gamma_1) - r(\gamma_2) - \epsilon(\gamma_2)$, which proves $s \in V_2(\epsilon(\gamma_2))$. Again, $v(\gamma_1|s) - v(\gamma_2|s) \leq -\ell(\gamma_1, \gamma_2)$, and the claims follows from:

$$\ell(\gamma_1, \gamma_2) - L(\alpha, \gamma_2) - L(\gamma_1, \alpha) > \ell(\gamma_1, \gamma_2) - L(\gamma_1, \alpha)$$
$$\quad + L(\gamma_2, \alpha) - z + \epsilon(\gamma_2)$$
$$\geq L(\gamma_1, \alpha) - L(\gamma_1, \alpha)$$
$$\quad - z + \epsilon(\gamma_2)$$
$$= -z + \epsilon(\gamma_2).$$

Now we show how Theorem 4.3.1 can be used to derive a sufficiency result. First, a definition.

Call Γ_g **two-cycle connected** if for every partition $\Gamma_1 \cup \Gamma_2 = \Gamma$, $\Gamma_1 \cap \Gamma_2 = \emptyset$, $\Gamma_1, \Gamma_2 \neq \emptyset$, there are $\alpha_1 \in \Gamma_1$ and $\alpha_2 \in \Gamma_2$ with $\ell(\alpha_1, \alpha_2) + \ell(\alpha_2, \alpha_1) = 0$.

Lemma 4.3.3 *Let g be IDS. If Γ_g is two-cycle connected, then g satisfies revenue equivalence.*

Proof. First, we show that if Γ_g is two-cycle connected, then any two nodes $\alpha, \beta \in \Gamma$ are connected in Γ_g by a finite path with nodes $\alpha = \alpha_0, \alpha_1, \ldots, \alpha_k = \beta$ such that $\ell(\alpha_i, \alpha_{i+1}) + \ell(\alpha_{i+1}, \alpha_i) = 0$ for $i = 0, \ldots, k-1$. Call such a path a zero-path. Suppose not. Then, there is a node $\alpha \in \Gamma$ that is not connected to all nodes in Γ by a zero-path with the described property. Define Γ_1 to be the set containing all nodes β that can be reached from α by a zero-path.

Let $\Gamma_2 = \Gamma \setminus \Gamma_1$. By assumption, $\Gamma_2 \neq \emptyset$. Because Γ_g is two-cycle connected, there are $\alpha_1 \in \Gamma_1$ and $\alpha_2 \in V_2$ with $\ell(\alpha_1, \alpha_2) + \ell(\alpha_2, \alpha_1) = 0$ contradicting the fact that $\alpha_2 \in \Gamma_2$.

Now, for any $\alpha, \beta \in \Gamma$, $L(\alpha, \beta)$ is bounded above by the length of a zero-path, $Z(\alpha, \beta)$ from α to β. Let $Z(\beta, \alpha)$ be the return path. Let $\ell(Z(\alpha, \beta))$ be the length of the zero path from α to β. Then $L(\alpha, \beta) \leq \ell(Z(\alpha, \beta))$ and $L(\beta, \alpha) \leq \ell(Z(\beta, \alpha))$. Therefore, $L(\alpha, \beta) + L(\beta, \alpha) \leq 0$. Because g is IDS, all cycles in Γ_g have nonnegative length. Therefore, $L(\alpha, \beta) + L(\beta, \alpha) = 0$. The theorem now follows from Theorem 4.3.1. ∎

We use Lemma 4.3.3 to derive revenue equivalence in a standard setting.

Theorem 4.3.4 *Let Γ be finite and T a (topologically) connected subset of $\mathbb{R}^k$. Each agent's valuation function $v_i(\gamma|t)$ is a continuous function in t for each $\gamma \in \Gamma$. Then every onto, IDS allocation rule $g: T^n \to \Gamma$ satisfies revenue equivalence.*

Proof. It is enough to show that Γ_g is two-cycle connected. We use the following fact from topology:

Let $T \subseteq \mathbb{R}^k$ be a connected set. Then any partition of T into subsets $T_1, T_2 \neq \emptyset$, $T_1 \cup T_2 = T$, $T_1 \cap T_2 = \emptyset$ satisfies $\overline{T}_1 \cap \overline{T}_2 \neq \emptyset$, where $\overline{T}_i$ is the closure of T_i in T.

Regard g as a function on T as before. Let $\Gamma_1 \cup \Gamma_2 = \Gamma$, $\Gamma_1 \cap \Gamma_2 = \emptyset$, $\Gamma_1, \Gamma_2 \neq \emptyset$ be a partition of Γ. Then $T = g^{-1}(\Gamma_1) \cup g^{-1}(\Gamma_2)$, $g^{-1}(\Gamma_1) \cap g^{-1}(\Gamma_2) = \emptyset$ is a partition of T, and $g^{-1}(\Gamma_1), g^{-1}(\Gamma_2) \neq \emptyset$, because g is onto the image of the relevant restriction of g. According to the fact above, there exists $t \in \overline{g^{-1}(\Gamma_1)} \cap \overline{g^{-1}(\Gamma_2)}$. Hence, there are sequences $(t_1^n) \subseteq g^{-1}(\Gamma_1)$ and $(t_2^n) \subseteq g^{-1}(\Gamma_2)$ with $\lim_{n \to \infty} t_1^n = \lim_{n \to \infty} t_2^n = t$. As Γ is finite, there must be $\alpha_1 \in \Gamma_1$ and $\alpha_2 \in A_2$ and subsequences $(t_1^{n_k}) \subseteq (t_1^n)$ and $(t_2^{n_m}) \subseteq (t_2^n)$ with $g(t_1^{n_k}) = \alpha_1$ for all k and $g(t_2^{n_m}) = \alpha_2$ for all m. Because v is continuous in the type,

$$0 = v(\alpha_2|t) - v(\alpha_1|t) - v(\alpha_2|t) + v(\alpha_1|t)$$
$$= \lim_{n \to \infty} (v(\alpha_2|t_2^{n_m}) - v(\alpha_1|t_2^{n_m}) + v(\alpha_1|t_1^{n_k}) - v(\alpha_2|t_1^{n_k})).$$

According to the definition of the arc length in Γ_g, the latter can be bounded from below as follows:

$$\lim_{n \to \infty} (v(\alpha_2|t_2^{n_m}) - v(\alpha_1|t_2^{n_m}) + v(\alpha_1|t_1^{n_k}) - v(\alpha_2|t_1^{n_k}))$$
$$\geq \ell(\alpha_1, \alpha_2) + \ell(\alpha_2, \alpha_1) \geq 0.$$

The last inequality is true because Γ_g has no negative cycles. Hence, all inequalities are equalities and $\ell(\alpha_1, \alpha_2) + \ell(\alpha_2, \alpha_1) = 0$. Consequently, Γ_g is two-cycle connected. ∎

The following example demonstrates that the continuity assumption cannot be relaxed.

4.3 Revenue Equivalence

Example 9 *Let there be one agent with type $t \in [0, 1]$ and two outcomes $A = \{a, b\}$. Let the agent's valuation be $v(a, t) = 1$, if $t < 1/2$ and $v(a, t) = 0$, if $t \geq 1/2$. Let $v(b, t) = 1/2$ for all t. That is, $v(a, \cdot)$ is discontinuous at $t = 1/2$. Let the allocation rule be the efficient one, that is, $f(t) = a$ for $t < 1/2$ and $f(t) = b$ otherwise. Then dominant strategy incentive compatibility is equivalent to $1 - \pi_a \geq 1/2 - \pi_b$ and $1/2 - \pi_b \geq -\pi_a$, which is satisfied whenever $|\pi_a - \pi_b| \leq 1/2$. For instance, $\pi_a = \pi_b = 0$ or $\pi'_a = 1/2$, $\pi'_b = 0$ are two payment schemes that make f truthful, but π and π' do not differ by a constant.*

4.3.1 A Demand-Rationing Example

In this section, an example for an economic setting is described where Theorem 4.3.1 can be used to identify revenue equivalence.

A supplier has one unit of a perfectly divisible good to be distributed among n retailers (agents). The type of agent i is his demand $t^i \in (0, 1]$. Given the reports $t \in (0, 1]^n$ of all agents, an allocation rule $g: (0, 1]^n \to [0, 1]^n$ assigns a fraction of the good to every agent such that $\sum_{i=1}^n g_i(t) \leq 1$. If an agent's demand is met, he incurs a disutility of 0, otherwise his disutility is linear in the amount of unmet demand. More precisely, agent i's valuation[19] if he is assigned quantity q_i is

$$v_i(q_i, t^i) = \begin{cases} 0, & \text{if } q_i \geq t^i, \\ q_i - t^i, & \text{if } q_i < t^i. \end{cases}$$

In this context, payments are reimbursements by the supplier for unmet demand.

Example 10 *We describe an implementable rule g that violates revenue equivalence. Call an allocation rule g dictatorial if there is an agent i that always gets precisely his demanded quantity, $g_i(t^i, t^{-i}) = t^i$ for all t^{-i}, and the other agents divide the remaining supply equally among themselves. We show that the dictatorial rule violates revenue equivalence.*

Let us pick agent 1 to be the dictator. Thus, $g_1(t) = t^1$ and $g_i(t) = (1 - t^1)/(n - 1)$ for all $i \neq 1$. First we show that g is IDS. Note that for agents $2, \ldots, n$, their assigned quantity does not depend on their report. Therefore, truthful reporting is a (weakly) dominant strategy for those agents. Consider the type graph for agent 1. It does not depend on the report of the other agents. For simplicity, we use v, t, and g instead of v^1, t^1, and g_1. Consider the type graph T_g. Let $s, t \in (0, 1]$ with $s < t$. We call (s, t) a forward arc and (t, s) a

---

[19] This valuation function appears in Holmstrom (1979) as an example to demonstrate that his smooth path-connectedness assumption cannot be weakened. The example can be used to show that the convexity assumption of the valuation function in Krishna and Maenner (2001) cannot be relaxed.

backward arc. Then

$$\ell(s, t) = v(g(t), t) - v(g(s), t) = v(t, t) - v(s, t) = t - s > 0, \text{ and}$$
$$\ell(t, s) = v(g(s), s) - v(g(t), s) = v(s, s) - v(t, s) = 0.$$

As all arcs have nonnegative length, there is no negative cycle and g is implementable. Furthermore, $L(t, s) = \ell(t, s) = 0$ for $t > s$.

To show that g violates revenue equivalence, we show that $L(s, t) = t - s > 0$. To that end, note that all paths from s to t that use only forward arcs have the same length, $t - s$. If a path from s to t contains a backward arc, the forward arcs of that path have a total length more than $t - s$. Hence, for all $s < t$, we have $L(s, t) + L(t, s) > 0$. Therefore, revenue equivalence does not hold according to Theorem 4.3.1. In fact, $P_1(t) = 0$ for all $t \in (0, 1]$ and $P_1(t) = t - 1$ for all $t \in (0, 1]$ are two payment schemes for agent 1 that make g dominant strategy incentive compatible. For the other agents, pick any constant payment scheme.

Example 10 shows that in this setting, not all implementable g satisfy revenue equivalence. A theorem describing sufficient conditions for *all* implementable g to satisfy revenue equivalence is necessarily silent here. Nevertheless, we can use Theorem 4.3.1 to identify properties of allocation rules that guarantee revenue equivalence in this setting. We state such properties formally in Theorem 4.3.5 below, and then show, for example, that the *proportional allocation rule* with $g_i(t) = t_i / \sum_{j=1}^n t_j$ satisfies these properties.

Theorem 4.3.5 *Let $g: (0, 1]^n \to [0, 1]^n$ be an allocation rule. If for every agent i and every report t^{-i} of the other agents, g satisfies any of the conditions below, then g is implementable and satisfies revenue equivalence.*

(1) $g_i(t^i, t^{-i})$ is continuous in t^i and $g_i(t^i, t^{-i}) < t^i$ for all $t^i \in (0, 1]$;
(2) $g_i(t^i, t^{-i})$ is continuous in t^i and $g_i(t^i, t^{-i}) > t^i$ for all $t^i \in (0, 1)$, $g_i(1, t^{-i}) = 1$;
(3) $g_i(t^i, t^{-i})$ is continuous and increasing in t^i and has exactly one fixed point $x \in (0, 1]$ with $g_i(x, t^{-i}) = x$, $g_i(t^i, t^{-i}) > t^i$ for all $t^i \in (0, x)$, $g_i(t^i, t^{-i}) < t^i$ for all $t^i \in (x, 1]$.

Proof. Fix agent i, t^{-i}, and use v, t, and g instead of v^i, t^i, and g_i. Furthermore, regard g as a function of i's type only. For all three cases, assume that g is continuous. Consider the corresponding type graph T_g.

(1) We have $g(t) < t$ for $t \in (0, 1]$. Let $s < t$. Using $g(s) < s < t$, the arc lengths in T_g are as follows:

$$\ell(s, t) = g(t) - g(s) \text{ and}$$

$$\ell(t, s) = \begin{cases} g(s) - g(t), & \text{if } g(t) < s, \\ g(s) - s, & \text{if } g(t) \geq s. \end{cases}$$

4.3 Revenue Equivalence

To verify implementability, we prove that any cycle in the type graph has nonnegative length. Note that for $s < t$, $\ell(s, t) = \ell(s, x) + \ell(x, t)$ for any $s < x < t$. We say that we *split* arc (s, t) at x if we replace (s, t) by arcs (s, x) and (x, t) in a path or cycle. Note that splitting forward arcs does not change the length of the path or cycle. Consider any finite cycle c with nodes c_1 to c_k and rename the nodes such that $c_1 < c_2 < \ldots < c_k$. Split every forward arc (c_u, c_v) with $u < v$ at all intermediate nodes $c_{u+1}, \ldots, c_{v-1}$ and call the resulting cycle c'; the cycle length remains the same. Consider some backward arc (c_v, c_u) with $u < v$. As c' is a cycle, it contains all forward arcs $(c_u, c_{u+1}), \ldots, (c_{v-1}, c_v)$. The length of the subcycle $(c_u, c_{u+1}, \ldots, c_{v-1}, c_v, c_u)$ is equal to

$$\ell(c_u, c_{u+1}) + \cdots + \ell(c_{v-1}, c_v) + \ell(c_v, c_u) = \ell(c_u, c_v) + \ell(c_v, c_u)$$

$$= \begin{cases} g(c_v) - c_u, & \text{if } g(c_v) \geq c_u, \\ 0, & \text{if } c_u > g(c_v) \end{cases}$$

and hence is nonnegative.

Removing the arcs of the subcycle $(c_u, c_{u+1}, \ldots, c_{v-1}, c_v, c_u)$ from c' leaves a new cycle c'' of smaller or equal length. As c'' is a cycle, we can repeat the argument finitely many times and terminate in the empty cycle with length 0. Hence, c has nonnegative length and g is implementable.

To verify that g satisfies revenue equivalence, we compute distances $L(s, t)$ and $L(t, s)$ for $s < t$. Note that $L(s, t) \leq \ell(s, t) = g(t) - g(s)$. We also claim that $L(t, s) \leq g(s) - g(t)$, so $L(s, t) + L(t, s) \leq 0$. Together with implementability, we get $L(s, t) + L(t, s) = 0$, and revenue equivalence holds by Theorem 4.3.1. To prove the claim, we consider two cases. If $g(t) < s$, $L(t, s) \leq \ell(t, s) = g(s) - g(t)$, and we are done. If $g(t) \geq s$, consider the sequence $(x_n)_{n=0}^\infty = (t, g(t), g(g(t)), \ldots)$. It is monotonically decreasing and bounded and so converges. The sequence $(g(x_n))_{n=0}^\infty = (g(t), g(g(t)), g^3(t), \ldots)$ converges for the same reason and $\lim_{n \to \infty} x_n = \lim_{n \to \infty} g(x_n) =: x$. As g is continuous, we have $x = g(x)$ if x is in $(0, 1]$. As there is no fixed point in $(0, 1]$, we conclude $x = 0$. Therefore, there exists some smallest index K such that $x_K < s$. Now consider the path π_K defined as $(t = x_0, \ldots, x_K, s)$. Path π_K has length $g^2(t) - g(t) + g^3(t) - g^2(t) + \cdots + g^{K+1}(t) - g^K(t) + g(s) - g^{K+1}(t) = g(s) - g(t)$. We conclude $L(t, s) \leq \ell(\pi_K) = g(s) - g(t)$, and we are done.

(2) We have $g(t) > t$ for $t \in (0, 1)$ and $g(1) = 1$. Let $s < t$. Using $g(t) \geq t > s$, the arc lengths in T_g can be computed as follows:

$$\ell(s, t) = \begin{cases} 0, & \text{if } g(s) \geq t; \\ t - g(s), & \text{if } g(s) < t \end{cases} \quad \text{and}$$

$$\ell(t, s) = 0.$$

As all arc lengths are nonnegative, all cycles have nonnegative length and g is implementable.

Since all paths have nonnegative length, we know that $L(t,s) = \ell(t,s) = 0$. We also claim that $L(s,t) = 0$, hence revenue equivalence holds. If $g(s) \geq t$, $L(s,t) = \ell(s,t) = 0$, and we are done. If $g(s) < t$, regard the sequence $(x_n)_{n=0}^\infty = (s, g(s), g(g(s)), \ldots)$. This sequence converges to the fixed point $x = 1$ by the same arguments as in (1). Let us regard paths π_k defined as $\pi_k = (s = x_0, \ldots, x_k, 1)$. As $\ell(x_i, g(x_i)) = 0$ for $i = 0, \ldots, k-1$ and $\ell(x_k, 1) = 1 - g(x_k)$, the length of π_k is $1 - g(x_k)$, which converges to 0 as k increases. Thus, $L(s, 1) = 0$. With $\ell(1, t) = 0$, it follows that $L(s, t) = 0$.

(3) We have that g is increasing on $(0, 1]$ and there is $x \in (0, 1)$ such that $g(x) = x$, $g(t) > t$ for $t \in (0, x)$, and $g(t) < t$ for $t \in (x, 1]$. There are six types of arcs whose lengths are as follows:

If $s < t \leq x$:	$\ell(s,t) = \begin{cases} 0, & \text{if } g(s) \geq t; \\ t - g(s), & \text{if } t > g(s); \end{cases}$	$\ell(t,s) = 0$;
If $s \leq x < t$:	$\ell(s,t) = g(t) - g(s)$;	$\ell(t,s) = 0$;
If $x < s < t$:	$\ell(s,t) = g(t) - g(s)$;	$\ell(t,s) = \begin{cases} g(s) - g(t), & \text{if } g(t) < s; \\ g(s) - s, & \text{if } s \leq g(t). \end{cases}$

Implementability can be verified in a manner similar to case (1). It can be checked that splitting forward arcs can only decrease the cycle length. When removing subcycles consisting of a backward arc (t, s), for $s < t$, together with the corresponding forward arcs, there are different cases. If $x < s < t$, the cycle length is nonnegative by arguments similar to case (1). If $s < x < t$ or $s < t < x$, the length of the backward arc is 0 and all forward arcs have nonnegative length. Thus, such a subcycle has nonnegative length, and the argument can be continued as in (1).

For revenue equivalence, we consider two cases. If $s < t \leq x$ or $x < s < t$, both s and t lie on a 0-length cycle just like in cases (1) and (2). Hence, $L(s,t) + L(t,s) = 0$. If $s \leq x < t$, an analogue argument as in (2) yields $L(s,x)) = 0$. With $\ell(x,t) = g(t) - g(x)$, we get $L(s,t) \leq g(t) - g(x)$. Similarly as in (1), we can show that $L(t,x) \leq g(x) - g(t)$. Consequently, and because $\ell(x,s) = 0$, we have $L(t,s) \leq L(t,x) + \ell(x,s) \leq g(x) - g(t)$. Thus, $L(s,t) + L(t,s) \leq 0$, and by implementability, $L(s,t) + L(t,s) = 0$. Hence revenue equivalence holds. ∎

COROLLARY 4.3.6 *For the demand-rationing problem, the proportional allocation rule g with $g_i(t) = t^i / \sum_{j=1}^n t^j$ for $i = 1, \ldots, n$ is implementable and satisfies revenue equivalence.*

Proof. For every agent i and every report of the other agents t_{-i}, the function $g_i(t^i, t^{-i}) = t_i / \sum_{j=1}^n t^j$ satisfies the assumptions of either case (1) or (3) in Theorem 4.3.5. ∎

4.4 THE CLASSICAL APPROACH

This section will outline the classical or Newtonian approach to characterizing incentive compatibility and its connections to the results described in the prior sections. For economy of exposition, we restrict attention to the case when valuations have a dot-product representation and the type space is convex.

One of the earliest and most well-known characterizations of IDS is in terms of the indirect utility function. A version of this result may be found in McAfee and McMillan (1988).

Let

$$V(s|t) = t \cdot g(s) - P(s).$$

In words, $V(s|t)$ is the utility from reporting type s given one is type t. Observe that $V(s|t)$ is an affine function of t. The indirect utility (or surplus) of an agent with type t is $U(t) = V(t|t)$.

Theorem 4.4.1 *g is IDS iff. $U(t)$ is convex, continuous, and $g(t)$ is a subgradient of U at t for all t in the relative interior of T.*

Proof. Suppose first that g is IDS. Then, convexity of U follows from the fact that a 'maximum' of affine functions is convex, that is,

$$U(t) = \sup_{s \in T} V(s|t).$$

From incentive compatibility, we know that

$$U(t) \geq V(s|t) = s \cdot g(s) - P(s) + g(s) \cdot (t - s)$$
$$= U(s) + g(s) \cdot (t - s). \tag{4.22}$$

Switching the roles of t and s in (4.22) yields

$$U(s) \geq U(t) + g(t) \cdot (s - t), \tag{4.23}$$

which is exactly the inequality that defines a subgradient of U at t. Combining (4.22) and (4.23) yields

$$g(t) \cdot (s - t) \leq U(s) - U(t) \leq g(s) \cdot (s - t). \tag{4.24}$$

Taking the limit as $s \to t$ in (4.24) establishes continuity of U.

The other direction is straightforward and omitted. ∎

Suppose in addition that U was differentiable everywhere (it need not be). Then, g would be the gradient of U. Specifically, $\nabla U(t) = g(t)$. Hence, U is the potential of the vector field g, and it follows that the line integral of a function with a potential is path independent. Thus, if $x : [0, 1] \to T$ is a piecewise smooth path such that $x(0) = s$ and $x(1) = t$,

$$U(t) = U(s) + \int g \cdot dx.$$

The integral here is a line integral along the path x. An immediate consequence of this observation is that the payoff depends only on the allocation rule (up to a constant) and revenue equivalence follows. Since differentiability of U cannot be guaranteed, additional conditions must be imposed to derive revenue equivalence along these lines.[20]

Assuming U and g are differentiable, convexity of U implies that the Jacobean of g, must be positive semidefinite. To see the connection to the 2-cycle condition, suppose that the allocation rule g can be represented by a matrix, B, that is, $v(g(t)|t) = Bt$. The 2-cycle condition in this case reduces to

$$(r - \tilde{r})' B (r - \tilde{r}) \geq 0 \qquad \forall r, \tilde{r} \in T,$$

where $'$ denotes "transpose." Note that

$$B = \frac{1}{2}(B + B') + \frac{1}{2}(B - B'),$$

that is, B can be decomposed into a symmetric part $\frac{1}{2}(B + B')$ and an anti-symmetric part $\frac{1}{2}(B - B')$. Further,

$$(r - \tilde{r})' B (r - \tilde{r}) = (r - \tilde{r})' \frac{1}{2}(B + B')(r - \tilde{r}).$$

Hence, if the symmetric part of B is positive semidefinite, the 2-cycle condition holds. Now, Theorem 24.9 from Rockafellar (1970) implies that g is IDS if and only if B is symmetric *and* positive semidefinite.[21] The 2-cycle condition implies only that the symmetric part of B is positive definite, not that B itself is symmetric. Therefore, the 2-cycle condition by itself is not sufficient to ensure that g is IDS. The observations above can be combined to produce the following characterization of IDS:[22]

Theorem 4.4.2 *Suppose T is convex and $v(\cdot|t)$ is a dot product valuation. Then, the following are equivalent.*

(1) g is IDS.
(2) g satisfies the 2-cycle condition and admits a potential (i.e., every line integral of g is path independent).

The reader should compare this theorem with Theorems 4.2.10 and 4.2.11.

The 2-cycle condition and path independence of the line integrals of g do not imply one another.

---

[20] See, for example, Krishna and Maenner (2001).
[21] Indeed, Rockafellar notes it as an exercise on p. 240. There is a close relationship between this result and characterizations of the producer's surplus function due to Hotelling (1932). Let Y be the production set and p the price vector. The producer chooses $y \in Y$ to maximize $p \cdot y$. Let $\Pi(p) = \max p \cdot y$ s.t. $y \in Y$ and $y(p)$ the optimal production choice for a given price vector p. Observe that $y(p)$ is a subgradient of Π at p and $\Pi(p)$ is convex in p. Further, the Jacobean of $y(p)$ is symmetric and positive definite. Notice that we could have replaced p by t, $\Pi(p)$ by the indirect utility function $U(t)$ and $y(p)$ by the allocation rule $g(t)$.
[22] See Jehiel and Moldovanu (2001) or Müller, Perea and Wolf (2007).

4.4 The Classical Approach

Example 11 *Here we exhibit an example that shows that the 2-cycle condition does not imply path independence. It will be convenient in this example to depart from the convention that a type is represented by the vector of valuations for each outcome. The type space is the simplex in $\mathbb{R}^3$. Let $x = (1, 0, 0)$, $y = (0, 1, 0)$ and $z = (0, 0, 1)$ be the extreme points of the type space. There are three outcomes, denoted a, b, and c. The allocation rule g is a linear mapping associating each type with a probability distribution over the three elementary outcomes. The outcome space Γ is the set of all possible probability distributions on $\{a, b, c\}$. Generic element $\gamma = (\gamma_a, \gamma_b, \gamma_c)$ indicates that a is achieved with probability γ_a, b with probability γ_b, and c with probability γ_c.*

Let V be the following matrix:

$$\begin{pmatrix} 2 & 0 & 3 \\ 3 & 2 & 0 \\ 0 & 3 & 2 \end{pmatrix}.$$

The allocation rule works as follows: $g(r) = Ir$, where I denotes the 3×3 identity matrix. For example, if the agent reports x as his type, then g awards him with the second-best outcome according to this type, that is, $g(x) = (1, 0, 0)$. Similarly, $g(y) = (0, 1, 0)$ and $g(z) = (0, 0, 1)$.

Define $v(g(r)|t) = r'Vt$. Thus, if the agent is of type x, his valuations for the three outcomes are given by the first column of V. The first element is his valuation for a, the second one his valuation for b, and the third one his valuation for c. Similarly, if the agent is of type y or z, his valuations for the elementary outcomes $\{a, b, c\}$ are given by the second and the third column of V, respectively.

As can be easily checked (by verifying that the symmetric part $\frac{1}{2}(V + V')$ of V is positive definite), g satisfies the 2-cycle inequality, that is, $(r - \tilde{r})' V (r - \tilde{r}) \geq 0$, $\forall r, \tilde{r} \in T$. Nevertheless, the 3-cycle $C = (x, y, z, x)$ has length $\ell(x, y) + \ell(y, z) + \ell(z, x) = -3$. The existence of a negative-length cycle implies that g is not IDS (see Theorem 4.2.1).

Consider now the path A consisting of the line segment between x and y, and path $\tilde{A}$ consisting of the line segment between x and z and the line segment between z and y. It is easy to verify that

$$\int_{x,A}^{y} g(r) = -\frac{3}{2} \quad \text{and} \quad \int_{x,\tilde{A}}^{y} g(r) = 3.$$

So the path integral of g from x to y is not independent of the path of integration.

Example 11 shows that the 2-cycle condition does not imply path independence. That the 2-cycle condition does not imply path independence depends on the assumption of multidimensional types. If we consider one-dimensional type spaces instead, the 2-cycle condition along with dot-product valuations would indeed imply path independence.

The next example shows that path independence does not imply the 2-cycle.

Example 12 *Consider the allocation of a single, indivisible object to an agent with type $t \in T = [0, 1]$ which reflects the value of the object to him. Given a report r of the agent, the allocation rule $g: T \mapsto [0, 1]$ assigns to him a probability for getting the object. Specifically, set $g(r) = -(2r - 1)^2 + 1$. The agent's valuation for the resulting allocation is $v(g(r) \mid t) = g(r)t$. Clearly, g is path independent but fails the 2-cycle condition.*

The other classical approach to the study of incentive compatibility is via the notion of cyclic monotonicity. An allocation rule g satisfies **cyclic monotonicity** if for every sequence of types $t^1, t^2, \ldots, t^k$ (with indices taken *mod* k), the following holds:

$$\sum_{i=0}^{k} [v(g(t^{i+1})|t^{i+1}) - v(g(t^{i+1})|t^i)] \geq 0.$$

Assuming a dot-product valuation, this becomes

$$\sum_{i=0}^{k} g(t^{i+1}) \cdot (t^{i+1} - t^i) \geq 0.$$

The historically accurate version of Rochet's Theorem is that g is IDS iff. g is cyclically monotone. The cyclic monotonicity condition and the absence of a negative cycle condition are identical.[23] To see it, reconsider incentive compatibility rewritten as follows:

$$U(t) - U(s) \leq g(t) \cdot (t - s).$$

We associate a network with this system with a vertex for each type in T and to each ordered pair (s, t) of types assign a directed edge from s to t of weight $w(s, t) = g(t) \cdot (t - s)$. The word 'weight' has been chosen intentionally to distinguish it from how length was defined earlier. Cyclic monotonicity corresponds to the absence of a negative "weight" cycle.

Clearly, $w(s, t) \neq \ell(s, t)$. However, as can be verified by direct computation, the length of a cycle $t^1 \to t^2 \to \cdots \to t^k \to t^1$ is equal to the weight of the cycle $t^k \to t^{k-1} \to \cdots \to t^1 \to t^k$. In words, the length of a cycle in the clockwise direction is equal to the weight of a cycle in the counterclockwise direction.

Whereas the node potentials associated with the "length" network correspond to incentive compatible transfers, the node potentials associated with the "weight" network correspond to the indirect utility or surplus obtained by the agents.

There is a close connection between the weight of a path and line integrals of g. Let L be some closed curve in T starting at $s \in T$ and ending at $t \in T$. Let

---

[23] This observation is made by Archer and Kleinberg (2008).

4.5 Interdependent Values

$x : [0, 1] \mapsto T$ be a piecewise smooth path such that $x(0) = s$ and $x(1) = t$. Consider the set of types $\{x(0), x(1/k), x(2/k), \ldots, x(\frac{k-1}{k}), x(1)\}$. The weight of the path that traverses $x(0), x(1/k), \ldots$ in that order is

$$\sum_{i=0}^{k-1} [x(\frac{i+1}{k}) - x(\frac{i}{k})] g(x(\frac{i+1}{k})). \quad (4.25)$$

The expression in (4.25) is precisely the right-hand Riemann sum of the line integral of g over L. Thus, cyclic monotonicity implies that every integral of g around a closed contour (a loop integral) is nonnegative. Reversing the orientation negates the integral, and we conclude that cyclic monotonicity implies that every loop integral of g is zero. Conversely, if every loop integral of g is zero and the Riemann sums converge from above, then g satisfies cyclic monotonicity. This observation and some refinements are summarized below. One new piece of notation is needed. Denote by $\Delta(t^1, t^2, t^3)$, the boundary of the triangle formed by the three points/types t^1, t^2, and t^3. Denote by $P(t^1, t^2, \ldots, t^k)$ the polygonal path through the cycle $t^1 \to t^2 \to \cdots \to t^k \to t^1$.

Theorem 4.4.3 *Suppose $T \subseteq \mathbb{R}^d$ is convex, $v : \Gamma \times T \mapsto \mathbb{R}$ is convex for each fixed $\gamma \in \Gamma$ and has nonempty sets of subgradients. Let $g : T \mapsto \Gamma$ satisfy the 2-cycle condition and decomposition monotonicity. Then, the following are equivalent:*

(1) g satisfies cyclic monotonicity.
(2) For every $t \in T$ and open neighborhood $O(t)$ around it

$$\int_{\Delta(t^1,t^2,t^3)} \nabla g(x) \cdot dx = 0$$

for all $t^1, t^2, t^3 \in O(t) \cap T$.
(3) For all $t^1, t^2, t^3 \in T$,

$$\int_{\Delta(t^1,t^2,t^3)} \nabla g(x) \cdot dx = 0.$$

(4) For all $t^1 \to t^2 \to \cdots \to t^k \to t^1$ with $k \geq 3$,

$$\int_{P(t^1,t^2,\ldots,t^k)} \nabla g(x) \cdot dx = 0.$$

This theorem is stated and proved in Berger, Müller, and Naeemi (2009). A special case had been derived earlier in Archer and Kleinberg (2008).

We conclude by noting that there is an extensive literature on monotone and cyclically monotone functions. See, for example, Rockafellar (1970) and chapter 12 of Rockafellar and Wets (1998).

4.5 INTERDEPENDENT VALUES

The interdependent values environment is designed to model situations where the private information of *other* agents, in addition to one's own, matters when determining the value of an outcome. In this environment, the type of an agent is often called a **signal**, and we do so here as well.

Let S be a set of signals. Denote by S^n the set of all n-agent profiles of signals.[24] An element of S^n will usually be written as $(s^1, s^2, \ldots, s^n)$ or $\mathbf{s}$. Let Γ be the set of outcomes or allocations.

The utility that agent i with signal $s^i \in S$ assigns to an allocation $\alpha \in \Gamma$ is denoted $v^i(\alpha|\mathbf{s})$. Observe that the utility agent i assigns to an allocation is function not just of i's signal, but the signals of all agents. Utilities are assumed quasilinear.

As before, a direct mechanism consists of two objects. The first is an **allocation rule**, which is a function

$$g: S^n \mapsto \Gamma.$$

The second is a **payment rule**, which is a function

$$P: S^n \mapsto \mathbb{R}^n,$$

that is, if the reported profile of signals is $(s^1, \ldots, s^n)$, agent i makes a payment of $P_i(s^1, \ldots, s^n)$.

Before continuing, we discuss two shortcomings of the setup. First, the valuation functions v^i are assumed to be common knowledge. Thus, it is only the signals that are private. In certain limited circumstances this is reasonable. For example, when the value is actually a statistical estimate of the actual value conditional on the signals, that is, a conditional expectation. If we know, for instance, the estimation procedure being used, we can deduce the functional form of the v^i's. The second is that a direct mechanism presumes a common language for communicating the signals. It is by no means clear that this is possible. For example, in some cases, agents may disagree on the interpretation of a signal.[25]

An allocation rule g is **ex-post incentive compatible** (EPIC) if there exists a payment rule P such that for all agents i and all signals $r \neq s^i$:

$$v^i(g(s^i, \mathbf{s}^{-i})|\mathbf{s}) - P_i(s^i, \mathbf{s}^{-i})$$
$$\geq v^i(g(r, \mathbf{s}^{-i})|\mathbf{s}) - P_i(r, \mathbf{s}^{-i}) \ \forall \ \mathbf{s}^{-i}. \quad (4.26)$$

It would be more correct to refer to g as being implementable in ex-post incentive compatible strategies.

The constraints (4.26) have a network interpretation. To see this, rewrite (4.26) as

$$P_i(s^i, \mathbf{s}^{-i}) - P_i(r, \mathbf{s}^{-i})$$
$$\leq v^i(g(s^i, \mathbf{s}^{-i})|\mathbf{s}) - v^i(g(r, \mathbf{s}^{-i})|\mathbf{s}) \ \forall \ \mathbf{s}^{-i}. \quad (4.27)$$

---

[24] The signal space does not need to be identical across agents. The assumption is for economy of notation only.

[25] These objections could apply to the independent private values case, but not when Γ is finite. When Γ is finite, the type or signal of an agent is a list of their values for each outcome. Thus, the designer does not need to know the map from types to valuations. It suffices for agents to report their valuations. Because types are valuations, there is no issue with interpreting them.

4.5 Interdependent Values

They coincide with the constraints of the dual of the shortest-path problem. Given this observation, a natural step is to associate with each i and $\mathbf{s}^{-i}$ a network with one node for each signal $s \in S$ and a directed arc between each ordered pair of nodes. Define the length of the arc directed from signal r to type q by

$$\ell(r, q|\mathbf{s}^{-i}) = v^i(g(q, \mathbf{s}^{-i})|\mathbf{s}^{-i}) - v^i(g(r, \mathbf{s}^{-i})|\mathbf{s}^{-i}).$$

Call this network $T_g(\mathbf{s}^{-i})$. Corollary 3.4.2 suggests that if this network has no negative-length cycles, then shortest-path lengths exist. Choose any one of the nodes as a source node and set $P_i(r, \mathbf{s}^{-i})$ equal to the length of the shortest path from the source to node t.[26]

It should be clear now that the prior techniques will apply in this setting to characterize ex-post incentive compatibility. However, the interdependent values setting introduces wrinkles absent in the independent case. Chief is the need to impose conditions on the behavior of v^i. To see why this is the case, consider the case of allocating a single indivisible object among two agents where signals are one-dimensional. Suppose agent 1 enjoys utility $a_{11}s^1 + a_{12}s^2$ from the good. Similarly, agent 2 enjoys utility $a_{21}s^1 + a_{22}s^2$ from consumption of the good. Assume that $a_{ij} \geq 0$ for all i, j and that the utility of either agent for not receiving the good is zero. Notice that the value of the good to an agent increases in her own signal as well as the other agent's signal.

Let g be the efficient allocation rule. Let r be a signal for agent 1 where, holding fixed the signal of agent 2, g does not allocate the good to agent 1. Then,

$$\ell(r, s^1|s^2) = a_{11}s^1 + a_{12}s^2$$

and

$$\ell(s^1, r|s^2) = -(a_{11}r + a_{12}s^2).$$

Therefore, the 2-cycle condition would imply that

$$0 \leq \ell(r, s^1|s^2) + \ell(s^1, r|s^2) = a_{11}(s^1 - r).$$

Hence, we obtain a familiar result: EPIC implies that the allocation rule must be monotone in signals.

Because the good is allocated to agent 1 when she has signal s^1, it follows that

$$a_{11}s^1 + a_{12}s^2 \geq a_{21}s^1 + a_{22}s^2 \Rightarrow (a_{11} - a_{21})s^1 \geq (a_{22} - a_{12})s^2.$$

Monotonicity requires this last inequality hold as we increase s^1. Hence, it follows that $a_{11} \geq a_{21}$. In words, agent 1's signal should have a larger impact on her utility than agent 2's signal.

---

[26] Strictly speaking, a direct application of Corollary 3.4.2 is not possible because it assumes a finite number of nodes. In the present case, the network has as many nodes as there are elements in S, which could be infinite. However, Theorem 3.4.4 allows us to do so.

Restricting attention only to the case of allocating a single object, the conclusions above are easily generalized. This we now do without proof (see Chung and Ely [2001]). Call the collection $(S, \{v^i\}_{i=1}^n)$ **uniformly ordered** if the following two conditions hold:

(1) The set S of signals is a partially ordered set with the order relation $\succeq$;
(2) For all $s^i, \hat{s}^i \in S$,
$$s^i \succ \hat{s}^i \iff v^i(s^i, \mathbf{s}^{-i}) > v^i(\hat{s}^i, \mathbf{s}^{-i}) \; \forall \mathbf{s}^{-i}$$

Theorem 4.5.1 *Let $(S, \{v^i\}_{i=1}^n)$ be uniformly ordered where v^i is the marginal value to agent i of the good. Then $g : S^n \mapsto \mathbb{R}$ is EPIC iff. g is monotone nondecreasing.*

The pair of valuation functions v^1, v^2 **cross at most once** if for any $s^1, \hat{s}^1, s^2, \hat{s}^2$ such that $s^1 \leq \hat{s}^1$ and $s^2 \leq \hat{s}^2$ we have that
$$v^1(s^1, s^2) > v^2(s^1, s^2) \Rightarrow v^1(\hat{s}^1, \hat{s}^2) \geq v^2(\hat{s}^1, \hat{s}^2).$$

Theorem 4.5.2 *Let $(S, \{v^i\}_{i=1}^2)$ be uniformly ordered and g be the efficient allocation rule for a single object. Then, g is EPIC iff. the pair (v^1, v^2) cross at most once.*

To generalize Theorem 4.5.2 to more than two agents, a stronger crossing condition is needed. Let $(S, \{v^i\}_{i=1}^n)$ be uniformly ordered and set
$$h^i(r) = v^i(r, \mathbf{s}^{-i}) - \max_{j \neq i} v^j(r, \mathbf{s}^{-i}).$$

The family $\{v^i\}_{i=1}^n$ satisfies **top single crossing** if for all i, h^i crosses zero at most once and from below.

Theorem 4.5.3 *Let $([0, 1], \{v^i\}_{i=1}^n)$ be uniformly ordered and g be the efficient allocation rule for a single object. If $\{v^i\}_{i=1}^n$ satisfies top single crossing, then g is EPIC.*

Notice that Theorem 4.5.3 only gives a sufficient condition.

When signals are multidimensional, the requirements on v^i imposed by both efficiency and EPIC can be extremely limiting. Consider again our simple example of allocating a single good when agents valuations are linear in the signals. The two cycle condition means that
$$a_{11} \cdot s^1 \geq a_{11} \cdot r.$$

Efficiency requires that
$$(a_{11} - a_{21}) \cdot s^1 \geq (a_{11} - a_{21}) \cdot r.$$

In other words, any pair of vectors s^1, r that satisfy the second inequality must satisfy the first. Unless the vector a_{11} is parallel to $a_{11} - a_{21}$, this will not be true. This observation implies the impossibility of implementing an efficient allocation in an ex-post incentive compatible way when signals are

4.6 BAYESIAN INCENTIVE COMPATIBILITY

The characterizations of dominant strategy incentive compatibility carry over, under the right conditions, to the case of Bayesian incentive compatibility.

Consider an agent i having true type t^i and reporting r^i whereas the others have true types $\mathbf{t}^{-i}$ and report $\mathbf{r}^{-i}$. The value that agent i assigns to the resulting allocation is denoted by $v^i\left(g\left(r^i, \mathbf{r}^{-i}\right) \mid t^i, \mathbf{t}^{-i}\right)$. Note that we allow for interdependence. Agents' types are independently distributed. Let π^i denote the density on T. The joint density π^{-i} on $\mathbf{t}^{-i}$ is then given by

$$\pi^{-i}(\mathbf{t}^{-i}) = \prod_{\substack{j \in N \\ j \neq i}} \pi^j(t^j).$$

Assume that agent i believes all other agents to report truthfully. If agent i has true type t^i, then his expected utility for making a report r^i is given by

$$U^i(r^i \mid t^i)$$
$$= \int_{T^{n-1}} \left(v^i\left(g\left(r^i, \mathbf{t}^{-i}\right) \mid t^i, \mathbf{t}^{-i}\right) - P_i\left(r^i, \mathbf{t}^{-i}\right)\right) \pi^{-i}\left(\mathbf{t}^{-i}\right) d\mathbf{t}^{-i}$$
$$= E_{-i}\left[v^i\left(g\left(r^i, \mathbf{t}^{-i}\right) \mid t^i, \mathbf{t}^{-i}\right) - P_i\left(r^i, \mathbf{t}^{-i}\right)\right]. \quad (4.28)$$

Assume $E_{-i}\left[v^i\left(g\left(r^i, \mathbf{t}^{-i}\right) \mid t^i, \mathbf{t}^{-i}\right)\right]$ to be finite $\forall r^i, t^i \in T$. An allocation rule g is **implementable in Bayes-Nash equilibrium** (IBN) if there exists a payment rule P such that $\forall i \in N$ and $\forall r^i, \tilde{r}^i \in T^i$:

$$E_{-i}\left[v^i\left(g\left(r^i, \mathbf{t}^{-i}\right) \mid r^i, \mathbf{t}^{-i}\right) - P_i\left(r^i, \mathbf{t}^{-i}\right)\right]$$
$$\geq E_{-i}\left[v^i\left(g\left(\tilde{r}^i, \mathbf{t}^{-i}\right) \mid r^i, \mathbf{t}^{-i}\right) - P_i\left(\tilde{r}^i, \mathbf{t}^{-i}\right)\right]. \quad (4.29)$$

An allocation rule g coupled with a payment rule that satisfies (4.29) is a **Bayes-Nash incentive compatible mechanism**.

Interchanging the roles of r^i and $\tilde{r}^i$ in (4.29) yields

$$E_{-i}\left[v^i\left(g\left(\tilde{r}^i, \mathbf{t}^{-i}\right) \mid \tilde{r}^i, \mathbf{t}^{-i}\right) - P_i\left(\tilde{r}^i, \mathbf{t}^{-i}\right)\right]$$
$$\geq E_{-i}\left[v^i\left(g\left(r^i, \mathbf{t}^{-i}\right) \mid \tilde{r}^i, \mathbf{t}^{-i}\right) - P_i\left(r^i, \mathbf{t}^{-i}\right)\right]. \quad (4.30)$$

Adding (4.29) and (4.30) yields the expected utility version of the 2-cycle condition. An allocation rule g satisfies the 2-cycle condition if $\forall i \in N$ and $\forall r^i, \tilde{r}^i \in T^i$:

$$E_{-i}\left[v^i\left(g\left(r^i, \mathbf{t}^{-i}\right) \mid r^i, \mathbf{t}^{-i}\right) - v^i\left(g\left(\tilde{r}^i, \mathbf{t}^{-i}\right) \mid r^i, \mathbf{t}^{-i}\right)\right]$$
$$\geq E_{-i}\left[v^i\left(g\left(r^i, \mathbf{t}^{-i}\right) \mid \tilde{r}^i, \mathbf{t}^{-i}\right) - v^i\left(g\left(\tilde{r}^i, \mathbf{t}^{-i}\right) \mid \tilde{r}^i, \mathbf{t}^{-i}\right)\right].$$

---

[27] Such impossibility results continue to hold with weaker notions of incentive compatibility. For responses to impossibility results of this kind, see Bikhchandani (2006) and Mezzetti (2004).

Obviously, the 2-cycle condition is necessary for IBN.

As before, the constraints in (4.29) have a natural network interpretation. For each agent i we build a complete directed graph T_g^i. A node is associated with each type and a directed arc is inserted between each ordered pair of nodes. For agent i, the length of an arc directed from r^i to $\tilde{r}^i$ is denoted $l^i(r^i, \tilde{r}^i)$ and is defined as:

$$l^i\left(r^i, \tilde{r}^i\right) = E_{-i}\left[v^i\left(g\left(r^i, \mathbf{t}^{-i}\right) \mid r^i, \mathbf{t}^{-i}\right) - v^i\left(g\left(\tilde{r}^i, \mathbf{t}^{-i}\right) \mid r^i, \mathbf{t}^{-i}\right)\right]. \tag{4.31}$$

Given our previous assumptions, the arc length is finite. For technical reasons, we allow for loops. However, note that an arc directed from r^i to r^i has length $l^i(r^i, r^i) = 0$.

The following is now straightforward.

Theorem 4.6.1 *An allocation rule g is IBN iff. there is no finite, negative length cycle in T_g^i, $\forall i \in N$.*

Suppose that Γ is finite. Then $g\left(t^i, \mathbf{t}^{-i}\right)$ is essentially a probability distribution over $\Delta(\Gamma)$. If valuations have a dot-product representation, that is,

$$v^i\left(g\left(\tilde{t}^i, \mathbf{t}^{-i}\right) \mid t^i, \mathbf{t}^{-i}\right) = t^i \cdot g\left(\tilde{t}^i, \mathbf{t}^{-i}\right).$$

Thus, Theorem 4.2.11 would apply with IDS replaced by IBN. If

$$E_{-i}\left[v^i\left(g\left(t^i, \mathbf{t}^{-i}\right) \mid t^i, \mathbf{t}^{-i}\right)\right],$$

were convex in t^i, then Theorem 4.2.12 would apply with IDS replaced by IBN.

CHAPTER 5

Efficiency

In this chapter we examine the allocation rule that chooses an efficient allocation. We use the machinery developed in Chapter 4 to show that an efficient allocation can be implemented in dominant strategies as well as determine the associated payments. Subsequently, we examine indirect implementations of the efficient allocation in the particular context of combinatorial auctions.

5.1 VICKREY-CLARKE-GROVES MECHANISM

Let T be the type space, n the number of agents, and Γ the set of allocations. The efficient allocation rule h is defined so that

$$h(t^1, t^2, \ldots, t^n) = \arg\max_{\gamma \in \Gamma} \sum_{i=1}^{n} t^i_\gamma.$$

It is assumed that T and Γ are such that the maximum exists. The efficient allocation rule is an example of an affine maximizer and so is IDS.

Theorem 5.1.1 h *is IDS.*

Proof. We use Rochet's Theorem. Fix the profile of types, $\mathbf{t}^{-i}$ of all agents except i. Observe that $h(t^i, \mathbf{t}^{-i}) = \alpha$ implies that

$$t^i_\alpha + \sum_{j \neq i} t^j_\alpha \geq t^i_\gamma + \sum_{j \neq i} t^j_\gamma \ \forall \gamma \in \Gamma.$$

Hence,

$$t^i_\alpha - t^i_\gamma \geq \sum_{j \neq i} t^j_\gamma - \sum_{j \neq i} t^j_\alpha \ \forall \gamma \in \Gamma.$$

Now consider a finite cycle of types in T_h. It will be enough to consider a cycle through s^i, r^i, and t^i where $h(s^i, \mathbf{t}^{-i}) = \gamma$, $h(r^i, \mathbf{t}^{-i}) = \beta$ and $h(t^i, \mathbf{t}^{-i}) = \alpha$. The argument for a cycle through four or more types is the same. Notice that

$$\ell(s^i, r^i) = r^i_\beta - r^i_\gamma \geq \sum_{j \neq i} t^j_\gamma - \sum_{j \neq i} t^j_\beta.$$

Similarly, $\ell(r^i, t^i) \geq \sum_{j \neq i} t_\beta^j - \sum_{j \neq i} t_\alpha^j$ and $\ell(t^i, s^i) \geq \sum_{j \neq i} t_\alpha^j - \sum_{j \neq i} t_\gamma^j$.
Hence,

$$\ell(s^i, r^i) + \ell(r^i, t^i) + \ell(t^i, s^i) \geq 0.$$

∎

When T is convex, we can mimic the argument in Lemma 4.2.9 combined with Theorem 4.3.1 to conclude that h satisfies revenue equivalence. With a bit more effort, one can relax the convexity of T assumption to smooth and piecewise connected.[1] Because the argument is similar to ones given in the previous chapter, it is omitted.

Using the allocation network, Γ_h, and Theorem 4.3.1 (assuming the appropriate conditions hold), we can determine the payments needed to yield incentive compatibility. Pick an $\alpha \in \Gamma$ and choose $P^\alpha(\mathbf{t}^{-i})$ arbitrarily. Set $P^\gamma(\mathbf{t}^{-i})$ to be $P^\alpha(\mathbf{t}^{-i})$ plus the length of the shortest path from α to γ in Γ_h. We now outline an argument that shows that the length of that shortest path is $\sum_{j \neq i} t_\gamma^j - \sum_{j \neq i} t_\alpha^j$.

Suppose the shortest path from α to γ is via β. Observe that (similar to the proof of Theorem 5.1.1)

$$\ell(\alpha, \beta) + \ell(\beta, \gamma) \geq \sum_{j \neq i} t_\alpha^j - \sum_{j \neq i} t_\beta^j + \sum_{j \neq i} t_\beta^j - \sum_{j \neq i} t_\gamma^j$$

$$= \sum_{j \neq i} t_\alpha^j - \sum_{j \neq i} t_\gamma^j \qquad (5.1)$$

and

$$\ell(\beta, \alpha) + \ell(\gamma, \beta) \geq \sum_{j \neq i} t_\beta^j - \sum_{j \neq i} t_\alpha^j + \sum_{j \neq i} t_\gamma^j - \sum_{j \neq i} t_\beta^j$$

$$= \sum_{j \neq i} t_\gamma^j - \sum_{j \neq i} t_\alpha^j. \qquad (5.2)$$

Adding (5.1) and (5.2) and invoking Theorem 4.3.1, we conclude

$$0 = \ell(\alpha, \beta) + \ell(\beta, \gamma) + \ell(\beta, \alpha) + \ell(\gamma, \beta) \geq 0.$$

Hence, equality must hold in (5.1) and (5.2).

Therefore, for all $\gamma \in \Gamma$, we have

$$P^\gamma(\mathbf{t}^{-i}) = \sum_{j \neq i} t_\alpha^j - \sum_{j \neq i} t_\gamma^j + P^\alpha(\mathbf{t}^{-i}).$$

It is usual to rewrite the payment formula in the following way:

$$P^\gamma(\mathbf{t}^{-i}) = -\sum_{j \neq i} t_\gamma^j + \phi(\mathbf{t}^{-i}),$$

---

[1] See, for example, Holmstrom (1979).

5.1 Vickrey-Clarke-Groves Mechanism

where ϕ is a function independent of t^i.

A particular mechanism associated with the efficient allocation is called the Vickrey-Clarke-Groves or VCG mechanism.[2] When types are nonnegative, the VCG mechanism specifies incentive-compatible transfers to satisfy individual rationality. Thus, in the VCG mechanism if $h(\mathbf{t}) = \gamma$, then for all i, $\phi(\mathbf{t}^{-i}) = \max_{\alpha \in \Gamma} \sum_{j \neq i} t_\alpha^j$

$$P^\gamma(\mathbf{t}^{-i}) = -\sum_{j \neq i} t_\gamma^j + \max_{\alpha \in \Gamma} \sum_{j \neq i} t_\alpha^j.$$

The surplus of agent i under the VCG mechanism when $h(\mathbf{t}) = \gamma$ is

$$t_\gamma^i - P^\gamma(\mathbf{t}^{-i}) = \sum_j t_\gamma^j - \max_{\alpha \in \Gamma} \sum_{j \neq i} t_\alpha^j.$$

It is more evocative to write it as

$$\max_{\theta \in \Gamma} \sum_j t_\theta^j - \max_{\alpha \in \Gamma} \sum_{j \neq i} t_\alpha^j.$$

This quantity is called agent i's **marginal product**.

The VCG mechanism returns an efficient allocation of resources that is individually rational by charging agents their marginal effect on the efficient choice of a social outcome. Such a payment scheme makes it a (weakly) dominant strategy for bidders to bid truthfully. Despite these attractive features, it has been argued that an indirect implementation of the VCG mechanism in the form of an ascending auction would be superior. For example, it has long been known that an ascending auction for the sale of a single object to bidders with independent private values exists that duplicates the outcomes of the VCG mechanism. It is the English clock auction, in which the auctioneer continuously raises the price for the object. An agent is to keep his hand raised until the price exceeds the amount the agent is willing to pay for the object. The price continues to rise only as long as at least two agents have raised hands. When the price stops, the remaining agent wins the object at that price.

In such an auction, no agent can benefit by lowering his hand prematurely, or by keeping his hand raised beyond the price he is willing to pay, as long as every other agent is behaving in a way consistent with some valuation for the object. The ascending auction, therefore, has the same incentive properties as its direct revelation counterpart.

There are at least two reasons why an ascending version of the VCG mechanism would be preferable to its direct revelation counterpart. The first is experimental evidence (Kagel and Levine 2001) that suggests that agents do not recognize that bidding truthfully is a dominant strategy in the direct version, but do behave according to the equilibrium prescription of the ascending counterpart. The second has to do with costs of identifying and communicating

---

[2] **V** for Vickrey (1961), **C** for Clarke (1971), and **G** for Groves (1973).

a bid. These costs have bite when what is being sold is a collection of heterogenous objects. To apply the VCG mechanism requires an agent to report their type for each possible allocation. In many cases this could be a prohibitively large number. An ascending version of the VCG mechanism auction avoids this expense in two ways. First, given an appropriate design, an agent only needs to decide if an allocations value exceeds some threshold at each stage. Second, agents need to focus only on the allocations that "matter" to them.[3]

The English clock auction is an instance of a Walrasian tâtonnement. The last word of the previous is sometimes rendered, in English, as "trial and error" or "groping." As imagined by Walras (1874), an auctioneer would announce prices for each good and service in the economy and agents would sincerely report their demand correspondences at the announced prices. By examining the balance between supply and reported demand, the auctioneer could adjust prices so as to match the two. In this way the auctioneer would, in Keynes' words grope, "like a peregrination through the catacombs with a guttering candle," her way to equilibrium prices. Here is what is known about tâtonnement processes.

(1) In general no tâtonnement process relying solely on the excess demand at the current price is guaranteed to converge to market clearing prices.
(2) Even when such a tâtonnement process exists, the prices can oscillate up and down.
(3) It is not necessarily in an agent's interest to sincerely report her demand correspondence at each price.

The allocation of a single object is remarkable because it is an exception on all three counts. In the remainder of this chapter we examine how far we may go beyond the sale of a single indivisible object in this regard. Specifically, we will derive ascending auctions that implement the Vickrey outcome for the sale of multiple heterogenous goods.

5.2 COMBINATORIAL AUCTIONS

An important class of environments for which many properties of the VCG mechanism have been derived are combinatorial auctions. There is a finite set of agents, N, and a finite set of indivisible objects (or goods), G. For each set of objects $S \subseteq G$, each agent $i \in N$ has a (monetary) nonnegative valuation $v_i(S) \geq 0$ (where $v_i(\emptyset) = 0$). To avoid minor technical issues, we will assume that each $v_i(S)$ is integral. We also assume that $v_i(S)$ is monotone, that is $S \subseteq T \Rightarrow v_i(S) \leq v_i(T)$. An agent i who consumes $S \subseteq G$ and makes a payment of $p \in \mathbb{R}$ receives a net payoff of $v_i(S) - p$.

A straightforward way to find an efficient allocation of the goods is via an integer linear program. This will have the advantage that the dual variables of the underlying linear programming relaxation can be interpreted as (possibly)

---

[3] Sometimes, in moving from a direct mechanism to an indirect mechanism, the set of equilibrium outcomes is enlarged. Restricting attention to ascending versions serves to limit this.

5.2 Combinatorial Auctions

supporting Walrasian prices.[4] More importantly, this will allow us to determine how rich the space of price functionals will need to be,

Let $y(S, i) = 1$ mean that the bundle $S \subseteq G$ is allocated to $i \in N$.[5]

$$V(N) = \max \sum_{j \in N} \sum_{S \subseteq G} v_j(S) y(S, j) \tag{5.3}$$

$$\text{s.t.} \quad \sum_{S \ni i} \sum_{j \in N} y(S, j) \leq 1 \quad \forall i \in G$$

$$\sum_{S \subseteq G} y(S, j) \leq 1 \quad \forall j \in N$$

$$y(S, j) = 0, 1 \quad \forall S \subseteq G, \forall j \in N.$$

The first constraint ensures that each object is assigned at most once. The second ensures that no bidder receives more than one subset. Call this formulation (CAP1). Denote the optimal objective function value of the linear programming relaxation of (CAP1) by $Z_1(N)$.

Even though the linear relaxation of (CAP1) might admit fractional solutions, let us write down its linear programming dual. To each constraint

$$\sum_{S \ni i} \sum_{j \in N} y(S, j) \leq 1 \quad \forall i \in G$$

associate the variable p_i, which will be interpreted as the price of object i. To each constraint

$$\sum_{S \subseteq G} y(S, j) \leq 1 \quad \forall j \in N$$

associate the variable π_j, which will be interpreted as the profit or surplus of bidder j. The dual is

$$Z_1(N) = \min \sum_{i \in G} p_i + \sum_{j \in N} \pi_j$$

$$\text{s.t.} \quad \pi_j + \sum_{i \in S} p_i \geq v_j(S) \quad \forall S \subseteq G$$

$$\pi_j \geq 0 \quad \forall j \in N$$

$$p_i \geq 0 \quad \forall i \in G$$

Let (π^*, p^*) be an optimal solution to this linear program. It is easy to see that

$$\pi_j^* = \max_{S \subseteq G} \left[v_j(S) - \sum_{i \in S} p_i^* \right]^+,$$

---

[4] The exposition here is based on de Vries, Schummer, and Vohra (2007).
[5] It is important to make clear the notation that, e.g., if agent i consumes the pair of distinct objects $g, h \in G$, then $y(\{g, h\}, i) = 1$, but $y(\{g\}, i) = 0$. An agent consumes exactly one set of objects.

which has a natural interpretation. Announce the price vector p^* and then let each agent choose the bundle that maximizes his payoff.

If $V(N) = Z_1(N)$, we can conclude that the efficient allocation is supported by single-item Walrasian prices on individual goods. Notice that the converse is also true. If the efficient allocation can be supported by single-item prices, then $V(N) = Z_1(N)$. One instance when $V(N) = Z_1(N)$ is when agents' preferences satisfy the (gross) substitutes property (see Section 5.5 for a definition).

In general, $V(N) \neq Z_1(N)$, so we cannot hope to support an efficient allocation using item prices. Suitably framed, we can interpret this as an impossibility result: There can be no ascending auction using only item prices guaranteed to deliver the efficient allocation. In this case, it is natural to look for a stronger formulation. A standard way to do this is by inserting additional variables. Such formulations are called extended formulations. Bikhchandani and Ostroy (2000) offer one.

To describe this extended formulation, let Π be the set of all possible partitions of the objects in the set G. If σ is an element of Π, we write $S \in \sigma$ to mean that the set $S \subset G$ is a part of the partition σ. Let $z_\sigma = 1$ if the partition σ is selected, and zero otherwise. These are the auxiliary variables. Using them we can reformulate (CAP1) as follows:

$$V(N) = \max \sum_{j \in N} \sum_{S \subseteq G} v_j(S) y(S, j)$$

$$\text{s.t.} \sum_{S \subseteq G} y(S, j) \leq 1 \quad \forall j \in N$$

$$\sum_{j \in N} y(S, j) \leq \sum_{\sigma \ni S} z_\sigma \quad \forall S \subseteq G$$

$$\sum_{\sigma \in \Pi} z_\sigma \leq 1$$

$$y(S, j) = 0, 1 \quad \forall S \subseteq G, \ j \in N$$

$$z_\sigma = 0, 1 \quad \forall \sigma \in \Pi.$$

Call this formulation (CAP2). In words, (CAP2) chooses a partition of G and then assigns the sets of the partition to bidders in such a way as to maximize efficiency. Let $Z_2(N)$ denote the optimal objective function value of the linear programming relaxation of (CAP2). Even though (CAP2) is stronger than (CAP1), it is still not the case that $V(N) = Z_2(N)$ always.

The dual of the linear relaxation of (CAP2) involves one variable for every constraint of the form:

$$\sum_{S \subseteq G} y(S, j) \leq 1 \ \forall j \in N;$$

call it π_j, which can be interpreted as the surplus that bidder j obtains. There will be a dual variable π_s associated with the constraint $\sum_{\sigma \in \Pi} z_\sigma \leq 1$, which

5.2 Combinatorial Auctions

can be interpreted to be the revenue of the seller. The dual also includes a variable for every constraint of the form:

$$\sum_{j \in N} y(S, j) \leq \sum_{\sigma \ni S} z_\sigma \quad \forall S \subseteq G,$$

which we denote $p(S)$ and interpret as the price of the bundle S.

The dual will be

$$Z_2(N) = \min \sum_{j \in N} \pi_j + \pi_s$$

$$\text{s.t.} \quad \pi_j \geq v_j(S) - p(S) \quad \forall j \in N, \ S \subseteq G$$

$$\pi_s \geq \sum_{S \in \sigma} p(S) \quad \forall \sigma \in \Pi$$

$$\pi_j, p(S), \pi_s \geq 0 \quad \forall j \in N, \ S \subseteq G$$

and has the obvious interpretation: minimizing the bidders' surplus plus seller's revenue. Notice that prices are now (possibly nonlinear) functions of the bundle consumed. Such bundle prices support the efficient allocation if and only if $V(N) = Z_2(N)$. Properly couched, we can interpret this as an impossibility result: In general, there can be no ascending auction using only nonlinear prices that is guaranteed to deliver the efficient allocation. One instance when $V(N) = Z_2(N)$ is when the v_j's are all submodular.[6]

To obtain an even stronger formulation (again due to Bikhchandani and Ostroy 2000), let μ denote both a partition of the set of objects *and* an assignment of the elements of the partition to bidders. Thus, μ and μ' can give rise to the same partition but to different assignments of the parts to bidders. Let Γ denote the set of all such partition-assignment pairs. We will write μ_j to mean that under μ, agent j receives the set μ_j. Let $\delta_\mu = 1$ if the partition-assignment pair $\mu \in \Gamma$ is selected, and zero otherwise. Using these new variables, the efficient allocation can be found by solving the formulation we will call (*CAP3*).

$$V(N) = \max \sum_{j \in N} \sum_{S \subseteq G} v_j(S) y(S, j)$$

$$\text{s.t.} \quad y(S, j) \leq \sum_{\mu : \mu_j = S} \delta_\mu \quad \forall j \in N, \ \forall S \subseteq G$$

$$\sum_{S \subseteq G} y(S, j) \leq 1 \quad \forall j \in N$$

$$\sum_{\mu \in \Gamma} \delta_\mu \leq 1$$

$$y(S, j) = 0, 1 \quad \forall S \subseteq G, \forall j \in N.$$

[6] See Parkes and Ungar (1999).

Bikhchandani and Ostroy (2000) showed that this formulation's linear programming relaxation has an optimal integer solution.

Theorem 5.2.1 *Problem (CAP3) has an optimal integer solution.*

Proof. We show that the linear relation of (CAP3) has an optimal integer solution. Let (y^*, δ^*) be an optimal fractional solution and $V_{LP}(N)$ its objective function value. To each partition-assignment pair μ, there is an associated integer solution: $y^\mu(S, j) = 1$ if S is assigned to agent j under μ, and zero otherwise.

With probability δ_μ^* select μ as the partition-assignment pair. The expected value of this solution is

$$\sum_\mu [\sum_{j \in N} \sum_{S \subseteq G} v_j(S) y^\mu(S, j)] \delta_\mu^*.$$

But this is equal to

$$\sum_{j \in N} \sum_{S \subseteq G} v_j(S) [\sum_{\mu : \mu_j = S} \delta_\mu^*] \geq \sum_{j \in N} \sum_{S \subseteq G} v_j(S) y^*(S, j) = V_{LP}(N).$$

The last inequality follows from

$$y^*(S, j) \leq \sum_{\mu : \mu_j = S} \delta_\mu^* \quad \forall j \in N, \forall S \subseteq G.$$

Thus, the expected value of the randomly generated integer solution is at least as large as the value of the optimal fractional solution. Therefore, at least one of these integer solutions must have a value at least as large as the optimal fractional solution. It cannot be strictly larger as this would contradict the optimality of (y^*, δ^*). So all y^μ have same objective function value as (y^*, δ^*) or there is no optimal fractional extreme point. ■

To write down the dual, we associate with each constraint $y(S, j) \leq \sum_{\mu : \mu_j = S} \delta_\mu$ a variable $p_j(S) \geq 0$ that can be interpreted as the price that agent j pays for the set S. To each constraint $\sum_{S \subseteq G} y(S, j) \leq 1$ we associate a variable $\pi_j \geq 0$ that can be interpreted as agent j's surplus. To the constraint $\sum_{\mu \in \Gamma} \delta_\mu \leq 1$ we associate the variable π_s that can be interpreted as the seller's revenue. The dual DP3 becomes

$$V(N) = \min \sum_{j \in N} \pi_j + \pi_s$$

$$\text{s.t.} \quad p_j(S) + \pi_j \geq v_j(S) \quad \forall j \in N, \forall S \subseteq G$$

$$- \sum_{\mu : \mu_j = S} p_j(S) + \pi_s \geq 0 \quad \forall \mu \in \Gamma$$

$$p_j(S) \geq 0 \quad \forall j \in N, \forall S \subseteq G$$

$$\pi_i \geq 0 \quad \forall i \in N \cup \{s\}.$$

5.3 The Core

Because (*CAP3*) has the integrality property, the dual above is exact. Hence, an efficient allocation can always be supported by Walrasian prices that are nonlinear and nonanonymous, meaning different agents see different prices for the same bundle.[7]

Lemma 5.2.2 *If* $\{\pi_i\}_{i \in N \cup s}$ *is an optimal solution to (DP3), then* $\pi_j \leq V(N) - V(N \setminus j)$ *for all* $j \in N$.

Proof. The optimal objective function value of problem (*CAP3*) is $V(N)$. Choose any $j \in N$. Now reduce the right-hand side of the $\sum_{S \subseteq G} y(S, j) \leq 1$ constraint to zero. The objective function value of this constrained program is $V(N \setminus j)$. Therefore, the dual variable associated with this constraint, π_j, at optimality, cannot exceed $V(N) - V(N \setminus j)$. ∎

5.3 THE CORE

One can associate with each instance of a combinatorial auction environment a cooperative game. Let $N^* = N \cup \{s\}$, where s denotes the seller, be the "players" of the cooperative game. For each $K \subseteq N^*$, set $u(K) = V(K \setminus \{s\})$ if $s \in K$, and zero otherwise. In words, only coalitions containing the seller generate value. The set of vectors $\pi \in \mathbb{R}^{|N^*|}$ that satisfy

$$\sum_{i \in N^*} \pi_i = u(N^*)$$

and

$$\sum_{i \in K} \pi_i \geq u(K) \ \forall K \subset N^*$$

is called the **core** of the cooperative game (N^*, u). The core in this case is nonempty because $\pi_s = u(N^*)$ and $\pi_i = 0$ for all $i \in N$ lies in the core.

Theorem 5.3.1 *Let* (π^*, p^*) *be an optimal solution to (DP3). Then* π^* *is an element of the core. If* π^* *is an element of the core, there exists* p^* *such that* (π^*, p^*) *is an optimal solution to (DP3).*

Proof. Suppose first that (π^*, p^*) is an optimal solution to (*DP3*). We show that π^* lies in the core. Clearly, the core constraints for sets K disjoint from $\{s\}$ are clearly satisfied. Next, for any $R \subseteq N$, $\{\pi_j^*\}_{j \in N}$, π_s^* and $\{p_j^*(S)\}_{j \in R}$ are a feasible dual solution to the primal program (*CAP3*) when restricted to the set R. Hence, by weak duality,

$$\sum_{j \in R} \pi_j^* + \pi_s^* \geq v(R) = u(R \cup s).$$

---

[7] One could ask whether it is possible to support the efficient allocation by Walrasian prices that can be decomposed into a nonlinear and nonanonymous part. The answer is yes – see Bikhchandani et al. (2002). There is also a connection between supporting price and communication complexity. See Nisan and Segal (2006).

Finally, by strong duality,

$$\sum_{j \in N} \pi_j^* + \pi_s^* = V(N) = u(N^*).$$

Now suppose π^* is an element of the core. Each element of the core is associated with some allocation of the goods designed to achieve π^*. Let μ^* be the partition assignment associated with π^*. If S is assigned to agent $j \in N$ under μ^*, set $p_j^*(S) = \max\{v_j(S) - \pi_j^*, 0\}$. It suffices to check that (π^*, p^*) is feasible in (DP3) to verify optimality because the core condition tells us that the objective function value of the solution is $V(N)$. ∎

In the context of combinatorial auctions, the marginal product of an agent j is $V(N) - V(N \setminus j)$. Consider now (CAP3). The dual variable associated with the second constraint, $\sum_{S \subseteq G} y(S, j) \leq 1$, can be interpreted as agent j's marginal product. To see why, for an agent $j \in N$, reduce the right-hand side of the corresponding constraint to zero. This has the effect of removing agent j from the problem. The resulting change in optimal objective function value will be agent j's marginal product. This argument shows only that among the set of optimal dual solutions to (CAP3), there is one that gives agent $j \in N$ her marginal product. Is there an optimal solution to (DP3) that simultaneously gives to each agent in N their marginal product? In general, no. But when the **agents are substitutes condition** (ASC) holds, the answer is yes. The definition of the ASC follows:

Agents Are Substitutes: For any subset of agents $M \subset N$,

$$V(N) - V(M) \geq \sum_{i \in N \setminus M} [V(N) - V(N \setminus i)] \tag{5.4}$$

The definition is due to Shapley and Shubik (1972).[8] The definition is easier to interpret if we replace M by $N \setminus K$ for some $K \subseteq N$. Then

$$V(N) - V(N \setminus K) \geq \sum_{i \in K} [V(N) - V(N \setminus i)].$$

In words, the marginal product of the set K is at least as large as the sum of the marginal products of each of its members. In a sense, the whole is greater than the sum of its parts.

Theorem 5.3.2 *If the ASC holds, there is a point π^* in the core of (N^*, u) such that $\pi_j^* = V(N) - V(N \setminus j)$ for all $j \in N$. Furthermore, π^* is the point in the core that minimizes π_s.*

---

[8] ASC is weaker than submodularity of V. Submodularity requires that for all M and K with $M \subseteq K \subseteq N$, $V(K) - V(M) \geq \sum_{i \in K \setminus M} [V(K) - V(K \setminus i)]$.

5.4 Ascending Auctions

Proof. If $\pi_j^* = V(N) - V(N \setminus j)$ for all $j \in N$ then, to be in the core, we need

$$\pi_s^* = V(N) - \sum_{j \in N}[V(N) - V(N \setminus j)].$$

For any $K \subseteq N$,

$$\sum_{j \in K} \pi_j^* + \pi_s^*$$

$$= \sum_{j \in K}[V(N) - V(N \setminus j)] + V(N) - \sum_{j \in N}[V(N) - V(N \setminus j)]$$

$$= V(N) - \sum_{j \in N \setminus K}[V(N) - V(N \setminus j)] \geq V(K) = u(K \cup s).$$

The last inequality follows from ASC.

Let π be any point in the core. Recall that

$$\sum_{j \in N} \pi_j + \pi_s = V(N)$$

and $\sum_{j \in N \setminus i} \pi_j + \pi_s \geq V(N \setminus i)$. Negating the last inequality and adding it to the previous equation yields $\pi_i \leq V(N) - V(N \setminus i)$. Because π^* achieves this upper bound simultaneously for all $j \in N$, it follows that π^* must have been chosen to be a point in the core that maximizes $\sum_{j \in N} \pi_j$. Equivalently, π^* is the point in the core that minimizes π_s. ∎

Thus, when the ASC holds, there exist supporting Walrasian prices that coincide with the prices charged under the VCG mechanism.

5.4 ASCENDING AUCTIONS

An ascending auction that produces the efficient allocation is an algorithm for finding the efficient allocation. By studying appropriate algorithms for finding the efficient allocation, we can derive ascending auctions. The idea that certain optimization algorithms can be interpreted as auctions goes back at least as far as Dantzig (1963). In particular, many of the ascending auctions proposed in the literature are primal-dual algorithms or subgradient algorithms for the underlying optimization problem for determining the efficient allocation. This observation yields a systematic way to derive an ascending implementation of the VCG mechanism. We describe both kinds of algorithms below.

5.4.1 Primal-Dual Algorithm

We begin with a high-level description of a primal-dual algorithm. Subsequently, it will be applied to (CAP3).

Consider the following linear program:

$$Z = \max cx$$
$$\text{s.t.} \sum_{j=1}^{n} a_{ij} x_j \leq b_i \quad \forall i = 1, \ldots, m$$
$$x_j \geq 0 \quad \forall j = 1, \ldots, n.$$

Here $A = \{a_{ij}\}$ is a constraint matrix of real numbers.
The dual to this linear program is:

$$Z = \min yb$$
$$\text{s.t.} \sum_{i=1}^{m} a_{ij} y_i \geq c_j \quad \forall j = 1, \ldots, n$$
$$y_i \geq 0 \quad \forall i = 1, \ldots, m.$$

Start with a feasible dual solution – say, y^*. Complementary slackness requires that

(1) $y_i^* > 0 \Rightarrow \sum_{j=1}^{n} a_{ij} x_j = b_i$;
(2) $\sum_{i=1}^{m} a_{ij} y_i^* > c_j \Rightarrow x_j = 0$.

To find a feasible solution to the primal that is complementary to y^*, we must solve:

$$\sum_{j=1}^{n} a_{ij} x_j = b_i \quad \forall i \text{ s.t. } y_i^* > 0$$

$$\sum_{j=1}^{n} a_{ij} x_j \leq b_i \quad \forall i \text{ s.t. } y_i^* = 0$$

$$x_j = 0 \; \forall j \text{ s.t. } \sum_{i=1}^{n} a_{ij} y_i^* > c_j$$

$$x_j \geq 0 \; \forall j \text{ s.t. } \sum_{i=1}^{n} a_{ij} y_i^* = c_j.$$

This system is called the restricted primal.

If the restricted primal is feasible, we are done. If not, by the Farkas Lemma, there is a solution to the alternative (the dual to the restricted primal), $\bar{y}$, say, such that

(1) $\bar{y} b < 0$, and
(2) $y^* + \epsilon \bar{y}$ is feasible in the dual for sufficiently small $\epsilon > 0$.

Choose as the new dual feasible solution $y^* + \epsilon \bar{y}$ and repeat the cycle.

5.4 Ascending Auctions

In the auction context, the dual variables will correspond to prices. The variable $\bar{y}$ corresponds to a direction of price adjustment. To ensure an "ascending" auction, one must guarantee that $\bar{y} \geq 0$.

A difficulty that arises in the auction context but not within the algorithm is the choice of ϵ. We finesse this by assuming, as stated earlier, that valuations are all integral. This allows us to raise prices a unit at a time without violating dual feasibility.

Primal-Dual Applied to (CAP3)

For any given list of prices $p_j(S) \geq 0$ (with $p_j(\emptyset) = 0$), the remaining variables of (DP3) can be chosen as follows to ensure feasibility. Throughout the algorithm, we maintain these equalities:

$$\pi_j = \max_{S \subseteq G}[v_j(S) - p_j(S)]$$

$$\pi_s = \max_{\mu \in \Gamma} \sum_{j \in N} p_j(\mu_j).$$

Interpret π_j as bidder j's potential surplus at current prices, and π_s as that of the seller.

Define, respectively, the set of *active bidders*, the *demand correspondence* of any bidder j, the seller's *supply correspondence*, and such assignments that do not assign nondemanded objects.

(1) $N^+ = \{j \in N : \pi_j > 0\}$
(2) $D_j(p) = \arg\max_{S \subseteq G}[v_j(S) - p_j(S)] = \{S \subseteq G : \pi_j = v_j(S) - p_j(S)\}$
(3) $\Gamma^* = \arg\max_{\mu \in \Gamma} \sum_{j \in N} p_j(\mu_j) = \{\mu \in \Gamma : \pi_s = \sum_{j \in N} p_j(\mu_j)\}$
(4) $\Gamma^*(D) = \{\mu \in \Gamma^* \mid \forall j \in N : \mu_j \in D_j(p) \cup \{\emptyset\}\}$.

The set $\Gamma^*(D)$ is the part of the seller's supply correspondence compatible with bidders' demand in the sense that no *non*demanded bundles are assigned.

The (nonredundant) complementary slackness (CS) conditions are:

$$j \in N^+ \implies \sum_{\emptyset \neq S \subseteq G} y(S, j) = 1$$

$$S \notin D_j(p) \cup \{\emptyset\} \implies y(S, j) = 0$$

$$\mu \notin \Gamma^* \implies \delta_\mu = 0.$$

Appending these constraints to (CAP3) yields the restricted primal:

$$y(S, j) = \sum_{\mu : \mu_j = S} \delta_\mu \quad \forall j \in N, \forall S \subseteq G$$

$$\sum_{\emptyset \neq S \subseteq G} y(S, j) = 1 \quad \forall j \in N^+$$

Efficiency

$$\sum_{\emptyset \neq S \subseteq G} y(S, j) \leq 1 \quad \forall j \notin N^+$$

$$\sum_{\mu \in \Gamma} \delta_\mu = 1$$

$$0 \leq y(S, j) \quad \forall j \in N, \forall S \in D_j(p) \cup \emptyset$$

$$y(S, j) = 0 \quad \forall j \in N, \forall S \notin D_j(p) \cup \emptyset$$

$$0 \leq \delta_\mu \quad \forall \mu \in \Gamma^*$$

$$\delta_\mu = 0 \quad \forall \mu \notin \Gamma^*.$$

These constraints further imply that for all $\mu \notin \Gamma^*(D)$, we have $\delta_\mu = 0$. We can simplify the program by removing all variables forced to zero. Henceforth, consider the variables $(\delta_\mu)_{\mu \notin \Gamma^*(D)}$ and $(y(S, j))_{S \notin D_j(p) \cup \{\emptyset\}}$ to be set to zero.

$$\text{(RP)} \quad y(S, j) = \sum_{\mu \in \Gamma^*(D): \mu_j = S} \delta_\mu \quad \forall j \in N, \forall S \in D_j(p) \cup \{\emptyset\}$$

$$\sum_{S \in D_j(p)} y(S, j) = 1 \quad \forall j \in N^+$$

$$\sum_{\emptyset \neq S \in D_j(p)} y(S, j) \leq 1 \quad \forall j \notin N^+$$

$$\sum_{\mu \in \Gamma^*(D)} \delta_\mu = 1$$

$$0 \leq y(S, j) \quad \forall j \in N, \forall S \in D_j(p) \cup \{\emptyset\}$$

$$0 \leq \delta_\mu \quad \forall \mu \in \Gamma^*(D).$$

If (RP) is feasible, then we began with an optimal solution (p, π), and are done. Otherwise, the Farkas Lemma implies that there exists a feasible solution to the following alternative system:

$$\text{(DRP)} \quad \lambda + \sum_{j \in N} \lambda_j < 0$$

$$\lambda_j + \rho_j(S) \geq 0 \quad \forall j \in N, \forall S \in D_j(p)$$

$$\rho_j(\emptyset) \geq 0 \quad \forall j \in N^+$$

$$\lambda - \sum_{j \in N} \rho_j(\mu_j) \geq 0 \quad \forall \mu \in \Gamma^*(D)$$

$$\rho_j(S) \gtrless 0 \quad \forall j \in N, \forall S \in D_j(p)$$

$$\lambda \gtrless 0$$

$$\lambda_j \gtrless 0 \quad \forall j \in N^+$$

$$\lambda_j \geq 0 \quad \forall j \notin N^+.$$

5.4 Ascending Auctions

We interpret $\rho_j(S)$ as a (direction of) price change for bundle S for bidder j. As a consequence, we do not change the prices of nondemanded bundles. We interpret λ_j as the change in agent j's surplus and λ as the change in the seller's surplus. When (RP) is infeasible, the first inequality of (DRP) states that total surplus must decrease.

To define an *ascending* auction, we search for a solution to (DRP) such that $\rho_j(S) \geq 0$ for all $j \in N$ and $S \in D_j(p) \cup \emptyset$. Such a solution may not exist if the variables for (DP3) (i.e., the $p_j(S)$'s) are chosen arbitrarily. It does, however, under a certain overdemand property defined below. That this property continues to hold throughout the adjustment procedure is what we intend to show.

To define it, we introduce artificial variables to (RP). For any $K \subseteq N^+$, define

$$(OD) \quad Z(K) = \max \ -\sum_{j \in K} z_j$$

$$\text{s.t.} \quad y(S, j) = \sum_{\mu \in \Gamma^*(D): \mu_j = S} \delta_\mu \quad \forall j \in N, \forall S \in D_j(p) \cup \{\emptyset\}$$

$$\sum_{S \in D_j(p)} y(S, j) + z_j = 1 \quad \forall j \in N^+$$

$$\sum_{\emptyset \neq S \in D_j(p)} y(S, j) \leq 1 \quad \forall j \notin N^+$$

$$\sum_{\mu \in \Gamma^*(D)} \delta_\mu = 1$$

$$0 \leq y(S, j) \quad \forall j \in N, \forall S \in D_j(p) \cup \{\emptyset\}$$

$$\delta_\mu \leq 1$$

$$0 \leq \delta_\mu \quad \forall \mu \in \Gamma^*(D)$$

$$0 \leq z_j \quad \forall j \in N^+.$$

In what follows, we make use of the fact that the feasible region of (OD) does not depend on the choice of K.

For a given list of (DP3) variables, say that **overdemand** holds if (OD) is feasible and $Z(N^+) < 0$. Feasibility of (OD) requires $\Gamma^*(D)$ to be nonempty. In turn, this implies that any unassigned object cannot be allocated in a way that creates additional revenue for the seller. In this sense, this prevents prices that get "too high." In addition, if overdemand holds, then (RP) is infeasible, because any feasible solution to (RP) would be a feasible solution to (OD) with objective function value $Z(N^+) = 0$.

The following definition is central to describing the price changes we use.[9] When overdemand holds, we say that a coalition $K \subseteq N^+$ is **undersupplied** if $Z(K) < 0$. The coalition is **minimally undersupplied** if for all $K' \subsetneq K$, $Z(K') = 0$. Notice that the definition does not change if "all $K' \subsetneq K$" were replaced with "all $K' = K \setminus \{j\}$ with $j \in K$." Furthermore, at least one nonempty, minimally undersupplied coalition must exist if overdemand holds.

The next theorem shows that nonnegative price changes can be chosen so that only minimally undersupplied bidders see positive price increases.

Theorem 5.4.1 *If overdemand holds, then for any minimally undersupplied coalition K, there is a solution to (DRP) such that $\rho_j(S) = 1$ for all $j \in K$ and $S \in D_j(p)$, and $\rho_j(S) = 0$ otherwise.*

Proof. Note that for all $T \subseteq N^+$,

$$Z(T) = \max_{\mu \in \Gamma^*(D)} |\{j \in T : \mu_j \in D_j(p)\}| - |T|.$$

Let $\Gamma^T = \arg\max_{\mu \in \Gamma^*(D)} |\{j \in T : \mu_j \in D_j(p)\}|$ denote the assignments that satisfy the greatest number of bidders in T.

Let K be minimally undersupplied. Because $Z(K \setminus j) = 0$ for any $j \in K$, it is easy to see that $Z(K) = -1$, that is $|K| - 1$ bidders can be satisfied. By the duality theorem of linear programming, we have that

$$Z(K) = \min \ \lambda + \sum_{j \in N} \lambda_j$$

$$\text{s.t.} \quad \lambda_j + \rho_j(S) \geq 0 \quad \forall j \in N, \forall S \in D_j$$

$$\rho_j(\emptyset) \geq 0 \quad \forall j \in N^+$$

$$\lambda - \sum_{j \in N} \rho_j(\mu_j) \geq 0 \quad \forall \mu \in \Gamma^*(D) \setminus \{\mu^0\}$$

$$\rho_j(S) \gtreqless 0 \quad \forall j \in N, \forall S \in D_j(p) \cup \{\emptyset\}$$

$$\lambda \gtreqless 0$$

$$\lambda_j \geq -1 \quad \forall j \in K$$

$$\lambda_j \geq 0 \quad \forall j \notin N \setminus K$$

$$\sigma \geq 0$$

Because $Z(K) < 0$, any optimal solution to the above program is a feasible solution to (DRP). We show that there is one where $\rho_j(S) = 1$ for all $j \in K$, and $S \in D_j(p)$, and $\rho_j(S) = 0$ otherwise.

---

[9] It is the natural generalization of Demange, Gale, and Sotomayor's (1986) concept of minimal overdemanded sets of objects.

5.4 Ascending Auctions

Set $\lambda_j = -1$ for all $j \in K$, and $\lambda_j = 0$ otherwise. Set $\lambda = |K| - 1$. This is clearly feasible. It is optimal because

$$\lambda + \sum_N \lambda_j = |K| - 1 - |K| = -1,$$

completing the proof. ∎

We have constructively defined the following **price adjustment** process.

(1) Identify a minimally undersupplied set of bidders, K.
(2) For each $j \in K$ and $S \in D_j(p)$, add $\rho_j(S) = 1$ to $p_j(S)$; otherwise do not change $p_j(S)$.
(3) For each $j \in K$, change π_j by $\lambda_j = -1$; for each $j \notin K$, do not change π_j.
(4) Increase π_s by $\lambda = |K| - 1$.

After such a price adjustment, it is clear that the demand correspondence for any bidder $j \notin K$ does not change. For $j \in K$, because valuations are assumed to be integral, a price increase of $\rho_j(S) = 1$ can only enlarge j's demand correspondence; no demanded bundle can become nondemanded (assuming integrality of prices throughout). Therefore, we have the following.

Lemma 5.4.2 *If prices are integral, then $D_j(\cdot)$ weakly increases after a price adjustment.*

Conversely, if $p_j(S)$ has increased during any price adjustment, S must be demanded by bidder j at all future iterations. Therefore, if the algorithm is initialized at zero prices ($p = 0$), then only demanded bundles can have positive prices.

Another observation is that, after a price adjustment, the seller's demand-compatible supply correspondence $\Gamma^*(D)$ can change in two ways. First, some $\mu \in \Gamma^*(D)$ could no longer be revenue maximizing after a price change. In this case, because the seller's revenue changes by $\lambda = |K| - 1$, the change in revenue from μ must be $|\{j \in K : \mu_j \in D_j(p)\}| \leq |K| - 2$. Second, some $\mu \notin \Gamma^*(D)$ could become revenue maximizing after a price change. This can happen only if $\{j \in K : \mu_j \in D_j(p)\} = K$ and before the price change,

$$\sum_{j \in N} p_j(\mu_j) = \sum_{j \in N} p_j(\mu'_j) - 1$$

where $\mu' \in \Gamma^K$.

To complete the proof that prices can only increase throughout the algorithm, we have the following result.

Theorem 5.4.3 *Beginning the algorithm at $p = 0$, overdemand holds after each iteration until termination.*

Proof. If each price is set to $p_j(S) = 0$, then overdemand holds: All bidders demand $D_j(p) = G$ and $\Gamma^*(D)$ is the set of n allocations where one bidder is assigned G and all other bidders receive nothing. Clearly, OD is feasible, $N^+ = N$, and $Z(N) < 0$.

Furthermore, if prices are integral at the beginning of an iteration, they remain integral after the iteration, by our specification of integral price increases.

To prove the result, it suffices to prove that feasibility of OD holds from one iteration to the next; if $Z(N^+) < 0$ after the iteration, then overdemand holds; otherwise, $Z(N^+) = 0$ and the algorithm terminates with an optimal solution. Below, the subscript or superscript t denotes the value of a variable during the tth iteration of the price adjustment.

Assume that OD is feasible during iteration t (and that prices are integral). We show that an optimal solution to OD in iteration t defines a feasible solution to OD in iteration $t + 1$. Let (δ^t, y^t, z^t) be such an optimal solution, and, without loss of generality, suppose it is integral,[10] so $\delta^t_{\hat{\mu}} = 1$ for some $\hat{\mu}$. The assignment $\hat{\mu}$ satisfies the demand of $|K^t| - 1$ bidders (where K^t is the minimally undersupplied set in round t). Because $\pi^{t+1}_s = \pi^t_s + |K^t| - 1$, that assignment is still revenue maximizing in round $t + 1$, i.e., $\hat{\mu} \in \Gamma^*_{t+1}$. By Lemma 5.4.2, this implies $\hat{\mu} \in \Gamma^*_{t+1}(D)$.

Therefore, a feasible solution to OD for iteration $t + 1$ can be obtained from (δ^t, y^t, z^t) by setting $\delta^{t+1}_{\hat{\mu}} = 1$ and all other δ^{t+1}_{μ} variables to zero. This implies $y(\hat{\mu}_j, j)^{t+1} = 1$ for all $j \in N$ such that $\hat{\mu}_j \in D_j(p) \cup \{\emptyset\}$, and $y(S, j)^{t+1} = 0$ otherwise. The z^{t+1}_j variables can obviously be chosen to complete a feasible solution; for all $j \in N^+$, $z_j = 1$ if and only if $\hat{\mu}_j = \emptyset$. ∎

Theorems 5.4.1 and 5.4.3 imply that prices are nondecreasing throughout the algorithm.

Theorem 5.4.4 *Modify the auction as follows: Fix an agent k. Whenever possible, choose at each stage a minimally undersupplied set that excludes k. Then at termination, agent k receives a payoff of $V(N) - V(N \setminus k)$.*

Proof. We show that at termination, $\pi_k = V(N) - V(N \setminus k)$. Suppose not. Because $0 \leq \pi_i \leq V(N) - V(N \setminus i)$ for all $i \in N$, it follows that $0 < \pi_k < V(N) - V(N \setminus k)$. Consider the last iteration, t, say, at which $\pi^t_k \geq V(N) - V(N \setminus k)$. Because prices rise a unit at a time, it must be that at this iteration, $\pi^t_k = V(N) - V(N \setminus k)$. By assumption, there was no minimally undersupplied set that excluded k. Therefore, N must be minimally undersupplied. In particular, there must be a $\mu \in \Gamma^*(D)^t$ such that $\mu_j \in D^t_j$ for all $j \neq k$. Furthermore, $\pi^t_s = \sum_j p^t_j(\mu_j)$ and $p^t_j(\mu_j) = v_j(\mu_j) - \pi^t_j$ for all $j \neq k$.

---

[10] Integrality of OD follows from integrality of (CAP3).

5.4 Ascending Auctions

Because $\pi_k^t = V(N) - V(N \setminus k)$ at this iteration, the values of the dual variables at this iteration are suboptimal. Therefore,

$$V(N) < \pi_s^t + \sum_{i \in N} \pi_i^t$$

$$= \sum_{j \in N} p_j^t(\mu_j) + \sum_{i \in N} \pi_i^t = \sum_{j \neq k}[v_j(\mu_j) - \pi_j^t] + \sum_{i \in N} \pi_i^t \leq V(N \setminus k) + \pi_k^t.$$

Therefore,

$$V(N) < V(N \setminus k) + [V(N) - V(N \setminus k)] = V(N),$$

which is a contradiction. ∎

Under sincere bidding, the primal-dual auction for (CAP3) terminates in prices that maximize the total surplus of the bidders. If the ASC holds, it follows by Theorem 5.3.2 that these prices correspond to Vickrey payments for the agents.

Theorem 5.4.5 *If the ASC holds, then, beginning at prices $p = 0$, the primal-dual auction for (CAP3) terminates in Vickrey prices.*

Proof. We show that at termination, $\pi_i = V(N) - V(N \setminus i)$ for all $i \in N$. Suppose not and consider the last instance of the auction when one or more bidders has a surplus that exceeds their marginal products. Call this set of bidders Q and suppose we are at iteration t. Then $\pi_j^t = V(N) - V(N \setminus j) + 1 \ \forall j \in Q$ and $\pi_j^t \leq V(N) - V(N \setminus j) \ \forall j \notin Q$.

At the end of iteration t, no bidder has a surplus that exceeds her marginal product. Therefore, all bidders in Q must have seen a unit price increase. Therefore, Q must be part of a minimally undersupplied set, K. Suppose first that Q is a strict subset of K. Then there is a $\mu \in \Gamma^*(D)$ at iteration t that satisfies all agents in Q. Let T be the set of agents satisfied in μ. Then $\pi_s^t = \sum_{j \in T} p_j^t(\mu_j)$ and $p_j^t(\mu_j) = v_j(\mu_j) - \pi_j^t$ for all $j \in T$. Because the auction has not terminated at iteration t, the current values of π_s^t, π_j^t are not optimal in (DP3). Therefore,

$$V(N) < \pi_s^t + \sum_{i \in N} \pi_i^t.$$

But

$$\pi_s^t + \sum_{i \in N} \pi_i^t = \sum_{j \in N} p_j^t(\mu_j) + \sum_{i \in N} \pi_i^t = \sum_{j \in T}[v_j(\mu_j) - \pi_j^t] + \sum_{i \in N} \pi_i^t$$

$$\leq V(T) + \sum_{i \notin T} \pi_i^t \leq V(T) + \sum_{i \notin T}[V(N) - V(N \setminus i)].$$

The last inequality follows from the fact that $Q \cap \{N \setminus T\} = \emptyset$. Therefore, $V(N) < V(T) + \sum_{i \notin T}[V(N) - V(N \setminus i)]$, which contradicts the ASC.

98 Efficiency

Now suppose $Q = K$. Because Q is minimally undersupplied, we can choose a $\mu \in \Gamma^*(D)$ at iteration t that satisfies all but one of the agents in Q, k. As before, let T be the set of agents satisfied by μ. As before,

$$V(N) < V(T) + \sum_{i \notin T} \pi_i^t \leq V(T)$$

$$+ \sum_{i \notin T \cup k} \pi_i^k + [V(N) - V(N \setminus k) + 1]. \tag{5.5}$$

For all $i \notin T \cup k$, we have $\pi_i^t \leq V(N) - V(N \setminus i)$. If we have strict inequality for any one of them, then inequality (5.5) yields a contradiction. Suppose not. Then $\pi_i^t = V(N) - V(N \setminus i)$ for all $i \notin Q = T \cup k$. Because no bidder outside Q sees a price increase, it means that at the end of the iteration, all bidders have a surplus of exactly their marginal product. If the algorithm terminates at this point, we are done. If not, there must be an undersupplied set at iteration $t + 1$. Choose any μ in the current $\Gamma^*(D)$. Then, as before,

$$\pi_s^{t+1} + \sum_{i \in N} \pi_i^{t+1}$$

$$= \sum_{j \in N} p_j^{t+1}(\mu_j) + \sum_{i \in N} \pi_i^{t+1} = \sum_{j \in T}[v_j(\mu_j) - \pi_j^{t+1}] + \sum_{i \in N} \pi_i^{t+1}$$

$$\leq V(T) + \sum_{i \notin T} \pi_i^t \leq V(T) + \sum_{i \notin T}[V(N) - V(N \setminus i)].$$

Therefore, $V(N) < V(T) + \sum_{i \notin T}[V(N) - V(N \setminus i)]$, which contradicts the ASC. ∎

5.4.2 Incentives

We now show that under certain conditions, sincere bidding is an *ex post Nash equilibrium* (defined below) in the primal-dual auction for (*CAP*3). That is, no bidder can benefit by bidding insincerely provided that all other bidders bid truthfully. This is a weaker equilibrium property than dominant strategies.[11] Ex post equilibrium requires that each bidder's strategy remains a best response, even after he learns the private information (valuations) of other bidders and hence their intended actions.

When Vickrey payments are implemented through the use of a direct revelation mechanism, bidders maximize their payoffs by bidding truthfully (i.e., truthfully reporting their valuations). Therefore, one hopes that an *ascending*

---

[11] In fact, dominant strategy equilibria are not obtainable in many reasonable dynamic-auction settings. It is beyond the scope of this book to elaborate further, but the reason for this is that *no* strategy can be optimal (best response) when another bidder is behaving in a way tailored to "punish" players using that given strategy; hence none can be dominant.
 Even the auctioneer could not detect such a bidder in a single run of the auction, because there are valuations that would be consistent with the bidder's observed behavior.

5.4 Ascending Auctions

auction (i.e., extensive form game) that implements Vickrey payments inherits good incentives properties. The argument for this follows the logic of the Revelation Principle. Suppose a bidder behaves in a way consistent with some false valuation function. Then he would receive elements and make payments that correspond to the Vickrey allocation/payments for the false valuation. By the dominant strategy incentive compatibility property of the sealed-bid mechanism, the bidder cannot be better off than if he had behaved truthfully.

Lemma 5.4.6 *Suppose the ASC holds and all bidders other than i bid truthfully at all times throughout the auction. If bidder i bids truthfully with respect to his true valuations v_i, then he receives a payoff at least as great as if he had bid in a way consistent with some false valuation $\tilde{v}_i$.*

Lemma 5.4.6 is insufficient to yield a general incentives result, however, because bidder i could bid in a way inconsistent with *any* valuation function. In certain special cases, one can show that if a bidder deviates from sincere bidding, there is a deviation consistent with some valuation function that yields the same payoff. In this case, one can obtain a general incentive result.[12]

5.4.3 Subgradient Algorithm

Another class of algorithms for linear programs that yield a natural auction interpretation are subgradient algorithms. We begin with a high-level description followed by an application to the allocation of heterogenous goods. In particular, we derive the i-bundle auction of Parkes and Ungar (1999), which is also the package bidding auction of Ausubel and Milgrom (2002). Consider the following linear program:

$$Z = \max cx$$
$$\text{s.t.} \sum_{j=1}^{n} a_{ij} x_j \leq b_i \quad \forall i = 1, \ldots, m$$
$$\sum_{j=1}^{n} b_{ij} x_j \leq d_i \quad \forall i = 1, \ldots, k$$
$$x_j \geq 0 \; \forall j = 1, \ldots, n$$

Here $A = \{a_{ij}\}$ and $B = \{b_{ij}\}$ are constraint matrices of real numbers. We now relax the first set of constraints (those associated with the matrix A):

$$Z(\theta) = \max cx + \sum_{i=1}^{m} \theta_i (b_i - \sum_{j=1}^{n} a_{ij} x_j)$$
$$\text{s.t.} \sum_{j=1}^{n} b_{ij} x_j \leq d_i \quad \forall i = 1, \ldots, k$$
$$x_j \geq 0 \quad \forall j = 1, \ldots, n$$

---
[12] See Bikhchandani, de Vries, Schummer, and Vohra (2008).

The multipliers associated with the relaxed constraints are the Lagrange multipliers. The duality theorem of linear programming implies that $Z = \min_{\theta \geq 0} Z(\theta)$. Finding the θ that solves $\min_{\theta \geq 0} Z(\theta)$ can be accomplished using the subgradient algorithm. For details, see II.3.6 of Nemhauser and Wolsey (1988).

Suppose the value of the Lagrange multiplier θ at iteration t is θ^t. Choose any subgradient of $Z(\theta^t)$ and call it s^t. Choose the Lagrange multiplier for iteration $t+1$ to be $\theta^t + \Delta_t s^t$, where Δ_t is a positive number called the step size. In fact, if x^t is an optimal solution associated with $Z(\theta^t)$,

$$\theta^{t+1} = \theta^t + \Delta_t(Ax^t - b).$$

Notice that $\theta_i^{t+1} > \theta_i^t$ for any i such that $\sum_j a_{ij} x_j^t > b_i$. The penalty term is increased on any constraint currently being violated. If $\sum_j a_{ij} x_j^t < b_i$, then the penalty term is *decreased*.

For an appropriate choice of step size at each iteration, this procedure can be shown to converge to the optimal solution. Specifically, $\Delta_t \to 0$ as $t \to \infty$ but $\sum_t \Delta_t$ diverges. Convergence of the subgradient algorithm is thus very slow.

The subgradient algorithm is guaranteed only to provide the optimal dual variables associated with the relaxed constraints at termination. The final primal "solution" need not be feasible. This is because each x^t is only guaranteed to be feasible for a relaxation of the original problem. However, in some applications it is possible to choose x^t such that $\lim_{t \to \infty} x^t$ is feasible for the original problem. This will be the case here.

Subgradient Algorithm Applied to (CAP3)

We apply the Lagrangean relaxation technique to formulation (CAP3):

$$V(N) = \max \sum_{j \in N} \sum_{S \subseteq M} v_j(S) y(S, j)$$

$$\text{s.t.} \quad y(S, j) \leq \sum_{\mu : \mu_j = S} \delta_\mu \quad \forall j \in N, \forall S \subseteq G$$

$$\sum_{S \subseteq G} y(S, j) \leq 1 \quad \forall j \in N$$

$$\sum_{\mu \in \Gamma} \delta_\mu \leq 1$$

$$\delta_\mu \geq 0 \quad \forall \mu \in \Gamma$$

$$y(S, j) = 0, 1 \quad \forall S \subseteq G, \forall j \in N.$$

5.4 Ascending Auctions

Relax the constraints $y(S, j) \leq \sum_{\mu:\mu_j=S} \delta_\mu$. Let $p_j(S) \geq 0$ be the corresponding multipliers. Then

$$V_p(N) = \max \sum_{j \in N} \sum_{S \subseteq G} [v_j(S) - p_j(S)]y(S, j) + \sum_{\mu \in \Gamma} \delta_\mu [\sum_j p_j(\mu_j)]$$

s.t. $\sum_{S \subseteq G} y(S, j) \leq 1 \quad \forall j \in N$

$\sum_{\mu \in \Gamma} \delta_\mu \leq 1$

$\delta_\mu \geq 0 \quad \forall \mu \in \Gamma$

$y(S, j) = 0, 1 \quad \forall S \subseteq G, \forall j \in N.$

Given $p_j(S)$, the solution to the relaxed problem can be found as follows:

(1) Choose $\delta_\mu = 1$ to maximize $\sum_{\mu \in \Gamma} \delta_\mu [\sum_j p_j(\mu_j)]$. In words, choose the solution that maximizes revenue at current prices. Call that solution the seller optimal allocation and label it μ^*.
(2) For each $j \in N$, choose a $B^j \in \arg\max_{S \subseteq G}[v_j(S) - p_j(S)]$ and set $y(B^j, j) = 1$. In words, each agent chooses a (possibly empty) utility-maximizing bundle at current prices.

From duality, $V(N) = \min_{p \geq 0} V_p(N)$. The subgradient algorithm solves the problem $\min_{p \geq 0} V_p(N)$.

Now apply the subgradient algorithm. If B^j is assigned to agent j under μ^*, the corresponding constraint

$$y(B^j, j) \leq \sum_{\mu:\mu_j = B^j} \delta_\mu$$

is satisfied at equality. So, in the subgradient iteration, the value of the multiplier $p_j(B^j)$ is unchanged. In words, if an agent's utility maximizing bundle is in μ^*, there is no price change on this bundle. If B^j is not assigned to agent j under μ^*, the corresponding constraint

$$y(B^j, j) \leq \sum_{\mu:\mu_j = B^j} \delta_\mu$$

is violated. So, by the subgradient iteration, the value of the multiplier $p_j(B^j)$ is increased by $\Delta > 0$. In words, if an agent's utility maximizing bundle is not in μ^*, that bundle sees a price increase. If μ^* assigns to agent j a bundle $S \neq B^j$, then the left-hand side of $y(S, j) \leq \sum_{\mu:\mu_j=S} \delta_\mu$ is zero whereas the right-hand side is 1. In this case, the value of the multiplier $p_j(S)$ is decreased by Δ. In words, a bundle that is a part of μ^* but not demanded sees a price decrease.

We have interpreted both the multipliers and the change in their value as the auctioneer setting and changing prices. We could just as well interpret the prices and price changes as bids and bid changes placed by the buyers. Thus, if a utility-maximizing bundle is part of a seller optimal allocation, the bid on it does not have to be increased. If a utility-maximizing bundle is not part of a seller optimal allocation, the bidder must increase her bid on it by exactly Δ.

The step size Δ of the subgradient algorithm becomes the bid increment. From the properties of the subgradient algorithm, for this auction to terminate in prices that support the efficient allocation, the bid increments must be very small and go to zero sufficiently slowly.

An Ascending Implementation of the Subgradient Algorithm

As described, the algorithm does not yield an auction with ascending prices. If one omits the step where prices are decreased, the algorithm is not guaranteed to terminate in an efficient allocation. To illustrate how a cavalier application of the algorithm can cause difficulties, consider an environment with two bidders $\{1, 2\}$ and two objects $\{a, b\}$. We assume that each bidder is interested in at most one of the two objects. Let $v_j(a)$ and $v_j(b)$ denote the value that agent j places on each object. We can formulate the problem of finding an efficient allocation as:

$$\max v_1(a)x_{1a} + v_1(b)x_{1b} + v_2(a)x_{2a} + v_2(b)x_{2b}$$

$$\text{s.t. } x_{1a} + x_{2a} \leq 1$$

$$x_{1b} + x_{2b} \leq 1$$

$$x_{1a} + x_{1b} \leq 1$$

$$x_{2a} + x_{2b} \leq 1$$

$$x_{1a}, x_{1b}, x_{2a}, x_{2b} \in \{0, 1\}.$$

Here $x_{ja} = 1$ if object a is assigned to bidder j. The linear relaxation of the above integer program has integral extreme points. So, we can relax the $x_{1a}, x_{1b}, x_{2a}, x_{2b} \in \{0, 1\}$ constraint to $0 \leq x_{1a}, x_{1b}, x_{2a}, x_{2b} \leq 1$.

Let us relax the first two constraints to form the following Lagrangean (with constant terms omitted from the objective function):

$$\max [v_1(a) - p_a]x_{1a} + [v_1(b) - p_b]x_{1b} + [v_2(a) - p_a]x_{2a}$$

$$+ [v_2(b) - p_b]x_{2b}$$

$$\text{s.t.} \quad x_{1a} + x_{1b} \leq 1$$

$$x_{2a} + x_{2b} \leq 1$$

$$0 \leq x_{1a}, x_{1b}, x_{2a}, x_{2b} \leq 1$$

We interpret the multipliers p_a and p_b as the price of a and b, respectively.

5.4 Ascending Auctions

Suppose $v_1(a) = v_2(a) = 2$ and $v_1(b) = v_2(b) = 1$. Set prices to zero to begin with. The optimal solution to the Lagrangean is $x_{1a} = x_{2a} = 1$. So, object a is overdemanded and object b not demanded at all. Increase the price of a by one unit and keep the price of b at 0 (which is what the subgradient algorithm dictates). At the new prices, there are multiple optimal solutions to the Lagrangean. However, the previous solution is still optimal. There is nothing in the description of the algorithm that prevents us from choosing the same solution as on the first iteration. If we do this, then we must raise the price of a, by 1/2, say.

When a is priced at 1.5 and b at zero, the optimal solution to the Lagrangean is $x_{1b} = x_{2b} = 1$. Now the price on b is raised. The pattern should now be clear. The price adjustments will diminish over time but it could be that at all times, all bidders demand the same good.

As described, the algorithm does not deliver an auction with ascending prices. However, a simple modification to the algorithm yields ascending prices without compromising its convergence properties.[13] The modified subgradient algorithm is described below:

(1) Initially set prices to zero, that is, $p_j(S) = 0$ for all $j \in N$ and $S \subseteq G$.
(2) For each $j \in N$, choose a $B^j \in \arg\max_{S \subseteq G}[v_j(S) - p_j(S)]$ and set $y(B^j, j) = 1$. In words, each agent chooses a utility-maximizing bundle at current prices.
(3) If a bidder is indifferent between participating and exiting the auction, he reports that fact. Call such a bidder inactive. All other agents are active.
(4) Given prices $p_j(S)$, choose $\delta_\mu = 1$ to maximize $\sum_{\mu \in \Gamma} \delta_\mu [\sum_j p_j(\mu_j)]$. In words, choose the solution that maximizes revenue at current prices. Call that solution the seller optimal allocation and label it μ^*.
(5) If for each active bidder k, B^k is assigned to agent k under μ^*, stop – a solution has been found.
(6) Otherwise, for each active bidder k, do:
   (a) If B^k is assigned to agent k under μ^*, there is no price change on this bundle.
   (b) If B^k is not assigned to agent k under μ^*, the value of the multiplier $p_k(B^k)$ is increased by $\Delta > 0$.
   (c) All other bundles maintain their current price.
(7) Go to 2.

Notice that all we have done is omit the price decrease step of the subgradient algorithm and change a price for only one agent at a time. Once an agent becomes inactive, he experiences no further price increases. One feature of the auction is that for any bidder, his surplus on any bundle is always nonnegative.

---
[13] This is essentially the modification of Kelso and Crawford (1982).

So, when a bidder becomes inactive, he is indifferent (at current prices) between all bundles and the emptyset. The algorithm just described is the i-bundle auction of Parkes and Ungar (1999), as well as the package bidding auction of Ausubel and Milgrom (2002).

Let p^t denote the vector of prices at iteration t. For the proof, we consider the lexicographic order on $\mathbb{N}^2$ where the first component corresponds to a bidder's current surplus and the second component to the size of his demand correspondence: that is, $(\lambda_i, |D_i(p)|) \leq (\lambda'_i, |D_i(p')|)$ if either $\lambda_i \leq \lambda'_i$ or $\lambda_i = \lambda'_i$ and $|D_i(p)| \leq |D_i(p')|$. For given p^t, consider the vector $h^t = ((\lambda_1, |D_1(p^t)|), (\lambda_2, |D_2(p^t)|), \ldots, (\lambda_n, |D_n(p^t)|))$ together with the order that is the cross product of $\mathbb{N}$-times the lexicographic order.

Lemma 5.4.7 *For each t, $h^t \geq h^{t+1}$. Further, if in the step from t to $t+1$ a price changed, then $h^t \neq h^{t+1}$.*

Proof. If no price for any k changed, clearly $h^t = h^{t+1}$.

Otherwise, only k's components are involved. By the price increase, the old B^k no longer provides the same surplus as before. If $|D_k(p^t)| > 1$, then there is at least one bundle in $D_k(p^t)$ of equal surplus that he can switch to (that is $\lambda^t_k = \lambda^{t+1}_k$ and $|D_k(p^t)| > |D_k(p^{t+1})|$). Otherwise, if $|D_k(p^t)| = 1$, k has no alternative of equal surplus, after B^k was knocked out, so $\lambda^t_k > \lambda^{t+1}_k$ holds. In both cases, we can conclude $h^t \gtrsim h^{t+1}$. ∎

Upon entering step 2 at time t, at least one conflict exists. Therefore, with the preceding lemma we obtain:

Lemma 5.4.8 *When step 3 is started at time t, then $h^t \gtrsim h^{t+n}$ for all $n \geq 1$ holds.*

The previous lemma shows that each agent's surplus diminishes or the size of his demand correspondance decreases from one round to the next. Because the size of the demand correspondence is finite and valuations are assumed to be integer, the auction must eventually terminate after a finite number of steps.

Lemma 5.4.9 *The auction terminates in the efficient allocation.*

Proof. Suppose the auction terminates in round T. Let σ be the terminal allocation and μ^* the efficient allocation. Let $\{S^1, S^2, \ldots, S^n\}$ be the bundles assigned to agents $1, 2 \ldots, n$ under μ^* and $\{K^1, K^2, \ldots, K^n\}$ the bundles assigned under σ. Then

$$\sum_{j \in N} v_j(S^j) > \sum_{j \in N} v_j(K^j).$$

Considering that μ^* was not a terminal allocation

$$\sum_{j \in N} p^T_j(S^j) < \sum_{j \in N} p^T_j(K^j).$$

5.5 GROSS SUBSTITUTES

Therefore,

$$\sum_{j \in N} [v_j(S^j) - p_j^T(S^j)] > \sum_{j \in N} [v_j(K^j) - p_j^T(K^j)].$$

Hence, there is at least one agent who has a utility-maximizing bundle that is not in σ at step T, contradicting termination. ∎

5.5 GROSS SUBSTITUTES

One instance of where the ASC holds is when bidders' preferences satisfy the gross substitutes property.[14] In the theory of discrete convexity, the substitutes property is known as $M^\#$-concavity.[15] Given object prices $p \in \mathbb{R}^{|G|}$, denote by $D(p)$ the collection of subsets of objects that maximize an agent's utility. In other words,

$$D(p) = \arg\max_{S \subseteq G} \{v(S) - \sum_{i \in S} p_i\}.$$

Gross Substitutes: For all price vectors p, p' such that $p' \geq p$, and all $S \in D(p)$, there exists $B \in D(p')$ such that $\{i \in S : p_i = p'_i\} \subset B$.

To prove that the ASC holds in this case, we make use of the fact that the linear relaxation of (5.3) has an integer solution. A proof of this fact can be found in Kelso and Crawford (1982) as well as Gul and Stachetti (1999).

Theorem 5.5.1 *If the value function of each agent satisfies the gross substitutes property, the ASC holds.*

Proof. For every subset of agents $M \subseteq N$ and object prices $p \in \mathbb{R}^{|G|}$, let

$$V_p(M) = \max \sum_{i \in M} \sum_{S \subseteq G} \left[v_i(S) - \sum_{g \in S} p_g \right] y(S, i) + \sum_{g \in G} p_g$$

$$\text{s.t.} \quad \sum_{S \subseteq G} y(S, i) \leq 1 \quad \forall i \in M$$

$$y(S, i) \geq 0 \quad \forall i \in M, \forall S \subseteq G.$$

Observe that $V_p(M)$ can be written more succinctly as

$$V_p(M) = \sum_{i \in M} \max_{S \subseteq G} \left[v_i(S) - \sum_{g \in S} p_g \right] + \sum_{g \in G} p_g.$$

By the duality theorem of linear programming,

$$V(M) = \min_{p \geq 0} V_p(M).$$

---

[14] First proved in Bikhchandani et al. (2002).
[15] The essential reference is Murota (2003). An abbreviated version of the same is Murota (2008).

Gul and Stachetti (1999) show that the set $\arg\min_{p \geq 0} V_p(M)$ forms a complete lattice, so a minimal element exists. Among all $p^* \in \arg\min_{p \geq 0} V_p(M)$, let p^M denote the minimal one.

It is also easy to see that if $M \subset M'$, then $p^M \leq p^{M'}$. Associated with each $V(M)$ is a partition of G. Denote by H_i^M the subset of objects that agent i receives in this partition, that is, $y(H_i^M, i) = 1$. For each $i \in N$, we have

$$V(N \setminus i) = V_{p^{N \setminus i}}(N \setminus i) \geq \sum_{j \in N \setminus i} \left[v_j(H_j^N) - \sum_{g \in H_j^N} p_g^{N \setminus i} \right] + \sum_{g \in G} p_g^{N \setminus i}$$

$$= \sum_{j \in N \setminus i} v_j(H_j^N) + \sum_{g \in H_i^N} p_g^{N \setminus i}$$

Hence,

$$V(N) - V(N \setminus i) \leq \sum_{j \in N} v_j(H_j^N) - \sum_{j \in N \setminus i} v_j(H_j^N) - \sum_{g \in H_i^N} p_g^{N \setminus i}$$

$$= v_i(H_i^N) - \sum_{g \in H_i^N} p_g^{N \setminus i}.$$

Let p' be a price vector such that

$$p'_g = \begin{cases} p_g^N & \text{if } g \in \bigcup_{i \in M} H_i^N \\ p_g^{N \setminus i} & \text{otherwise, where } g \in H_i^N, i \in N \setminus M. \end{cases}$$

Then, since $p' \geq p^M$,

$$V(M) = V_{p^M}(M) \leq V_{p'}(M) = \sum_{i \in M} \max_{H \subseteq G} \left[v_i(H) - \sum_{g \in H} p'_g \right] + \sum_{g \in G} p'_g$$

$$= \sum_{i \in M} v_i(H_i^N) + \sum_{j \in N \setminus M} \sum_{g \in H_j^N} p_g^{N \setminus j}.$$

Therefore,

$$V(N) - V(M) \geq \sum_{i \in N \setminus M} v_i(H_i^N) - \sum_{i \in N \setminus M} \sum_{g \in H_i^N} p_g^{N \setminus i}$$

$$\geq \sum_{i \in N \setminus M} [V(N) - V(N \setminus i)],$$

and the condition holds. ∎

The theorem is false if the valuations are merely submodular (see, for example, Gul and Stachetti 1999).

5.6 AN IMPOSSIBILITY

The ASC, implied by the gross substitutes condition on valuation functions, is sufficient for the existence of an ascending auction that implements the VCG outcome. We now argue that the gross substitutes condition is somewhat necessary for this existence, in the following sense. If one bidder's valuation function fails the gross substitutes condition at valuations which may effect other bidders' VCG payments, then there exists a family of perturbed valuation functions for the other bidders (which *does* satisfy gross substitutes) over which no ascending auction (defined below) can always implement VCG payments.[16] The intuition behind the result below can be given with a simple example. Suppose three bidders have the following valuations for $G = \{a, b\}$.

Bidder	Bundle		
	a	b	ab
1	2	3	8
2	1	5	6
3	$4+\epsilon$	$1+\epsilon$	$4+\epsilon$

We restrict attention to the cases $\epsilon \in [-.5, .5]$, in which case it is efficient to give b to Bidder 2 and a to Bidder 3. Their respective VCG payments are $4 - \epsilon$ and 3 (whereas Bidder 1 pays and receives nothing).

If an ascending auction uses only "real price" information to determine allocations and payments (as we define below), then it must do two things. First, it must conclude by offering good a to Bidder 3 at a price of 3. Second, it must determine the value of ϵ in order to offer good b to Bidder 2 at the correct price. However, for the class of Bidder 3's valuation functions obtained by varying $\epsilon \in [-.5, .5]$, the value of ϵ cannot be inferred from his demand set until Bidder 3 "demands" the empty set. To see why, suppose at price vector p for agent 3, she prefers $\{a\}$ to $\{b\}$. Then

$$4 + \epsilon - p_a \geq 1 + \epsilon - p_b.$$

Notice that the ϵ cancels, so no information about ϵ can be recovered from this comparison. The same is true of other pairs of bundles. However, when Bidder 3 demands the empty set, then information about ϵ is revealed. This happens only when her price for a exceeds $4 + \epsilon > 3$. This contradicts the fact that her price for that objects ascends throughout the auction and ends at 3.

This idea can be extended to any situation where one bidder has a valuation function over two objects that does not satisfy gross substitutes. To formalize this, we define an ascending auction.

A **price path** is a function $p: [0, 1] \to \mathbb{R}^{2^G \times N}$. For each bundle of goods $H \subseteq G$, interpret $p_{i,H}(t)$ to be the price seen by bidder i for bundle H, at "time" t.

---

[16] Gul and Stachetti (2000) prove a result of a similar flavor when auction mechanisms are required to assign additive, anonymous prices to objects and bundles.

An **ascending auction**, A, assigns to each profile of bidder valuation functions, $v \in \mathbb{R}_+^{2^G \times N}$, a price path $A(v) = p$ that depends only on demand: For all valuation profiles $v, v' \in \mathbb{R}_+^{2^G \times N}$, letting $p = A(v)$,

$$\text{if } [\forall t \in [0, 1], i \in N, D_i(p(t); v) = D_i(p(t); v')]$$

$$\text{then } A(v') = A(v). \tag{5.6}$$

That is, if one profile v yields a particular price path p under an ascending auction, and if, under another profile v', each bidder would demand the same objects at the prices in path p, then the profile v' yields the same path p in the auction. Information is revealed only through demand in the auction.[17]

An ascending auction then assigns goods such that each bidder i receives a bundle, H, in his demand set at prices $p(1)$, and charges that bidder $p_{i,H}(1)$. In this sense, we require prices in an auction not to be merely artificial constructs; they represent potential payments.

Unfortunately, when at least one bidder has a valuation function that does not satisfy the gross substitutes condition, such an auction must fail to always implement VCG outcomes. We prove this formally for the case of two objects and three bidders. Similar results could be obtained for other cases at the expense of additional notation.

Theorem 5.6.1 *Suppose $G = \{a, b\}$ and $|N| \geq 3$. Suppose one bidder's valuation function fails the gross substitutes condition. Then there exists a class of gross substitutes valuation functions for the other bidders such that no ascending auction implements the VCG outcome for each profile from that class.*

Proof. To prove the result, suppose, without loss of generality, that Bidder 1's valuations violates gross substitutes, and $v_1(a) = 1$. Denote $v_1(b) = x$, and $v_1(ab) = 1 + x + y$ where $y > 0$ (failing gross substitutes). Let Bidder 2 have valuations $v_2(a) = 1$, $v_2(b) = x + y$, and $v_2(ab) = x + y$. For $\epsilon \in [-.5, .5]$, consider Bidder 3's valuations to be of the form $v_3(a) = 1 + y + \epsilon$, $v_3(b) = 1 + \epsilon$, and $v_3(ab) = 1 + y + \epsilon$. Ignore additional bidders (or assign them infinitesimal valuations).

When $\epsilon \in [-.5, .5]$, the efficient allocation assigns good b to Bidder 2 and good a to Bidder 3. The VCG payments for those two bidders are $x - \epsilon$ and 1, respectively (whereas Bidder 1 obviously pays nothing).

When, for example, $\epsilon = 0$, the price path must finish by yielding $p_{a,3}(1) = 1$. Because $p_{a,3}(\cdot) \leq 1$ throughout the auction (by monotonicity), Bidder 3 never demands the empty set. Hence, his demand sets would be the same for any other value of $\epsilon \in [-.5, .5]$.

By the requirement (5.6) of an ascending auction, this means that the price paths must be constant with respect to ϵ. This contradicts the fact that it should yield Bidder 2's VCG payment, which depends on ϵ. ∎

---

[17] Naturally, the empty set may be demanded, so information may be revealed when prices get "too high" for a bidder.

5.7 A RECIPE

Here is a simple recipe for deriving ascending auctions:

(1) First, verify that the ASC holds.
(2) Formulate as a linear program the problem of finding the efficient allocation. Appropriate variables in the dual can be interpreted as the prices paid by buyers as well as the surplus they receive. The formulation must be sufficiently rich that the VCG payments are contained in the set of feasible dual solutions. Furthermore, VCG payments must correspond to an *optimal* dual solution.
(3) We have a primal problem whose optimal solutions correspond to efficient allocations. Given (1) and (2), we know that among the set of optimal dual solutions that support the efficient allocation is the one that corresponds to the VCG payments. In fact, these will be the prices that minimize the surplus of the seller. Call this the buyer optimal dual solution.
(4) Choose either a primal-dual or subgradient algorithm for this linear program that terminates in the buyer optimal dual solution.
(5) Interpret the chosen algorithm as an auction. This is easy given the structure of such algorithms. They operate by choosing at each iteration a feasible dual solution (i.e., prices) and then forcing a primal solution that is complementary to the current dual solution. This corresponds to bidders reporting their demands at current prices. If a primal solution complementary to the current dual solution exists, the algorithm stops. This corresponds to all demands being satisfied at current prices. If no such primal solution exists, the dual variables must be adjusted. Nonexistence of a complementary primal solution corresponds to an imbalance between supply and demand. The price change attempts to adjust the imbalance.

The algorithms differ in two ways. First, a primal-dual algorithm requires a bidder's entire demand correspondence at each iteration. A subgradient algorithm requires only an element of the demand correspondence. Second, they differ in how prices are adjusted. The virtue of the primal dual-based algorithm over the subgradient algorithm is in convergence speed.
(6) Given the algorithm terminates with the efficient allocation and charges each buyer their VCG payment, it follows immediately that bidding sincerely is an ex-post Nash equilibrium of the auction.

CHAPTER 6

Revenue Maximization

This chapter is concerned with techniques for determining a Bayesian incentive compatible mechanism that maximizes the expected revenue of the designer. On occasion, these mechanisms will actually be dominant strategy incentive compatible. In contrast to much of the literature, we assume that the type space is discrete. We know of no modeling reason to prefer a continuous type space to a discrete one. Rather, the choice should be based on analytical tractability. The discrete type space assumption allows one to formulate the problem of finding a revenue-maximizing Bayesian incentive compatible mechanism as a linear program. It will be seen that the solution for discrete type case is the natural analog of the solution for the continuous case. Indeed, the solution in the continuous type case can be obtained by discretizing the type space and taking the limit. We use this technique to derive a solution for an instance with budget constraints for continuous types – one for which no solution was previously known.

Let Γ be the set of feasible allocations of the resources among the agents and the designer. Let T be a finite set of an agent's types (possibly multidimensional) and N the set of agents. As before, the type of agent k is denoted t^k. Let T^n be the set of all profiles of types where $n = |N|$. A profile in which attention is to be focused on agent k will frequently be written as $(t^k, \mathbf{t}^{-k})$. If $\gamma \in \Gamma$, then agent k with type t^k assigns monetary value $v(\gamma | t^k)$ to the allocation $\gamma \in \Gamma$.

Assume that types are selected independently from T according to a common distribution. Denote by f_t the probability that an agent k has type $t \in T$, that is, $t^k = t$. For economy of notation, we will assume that $f_t > 0$ for all $t \in T$. Denote the probability of a profile $\mathbf{t}^{-k} \in T^{n-1}$ being realized by $\pi(\mathbf{t}^{-k})$.

Let P_{t^k} be the expected payment that an agent who announces type t^k makes. An allocation rule assigns to each member of T^n an element of Γ.[1] If α is an allocation rule, write $\alpha_k[t^k, \mathbf{t}^{-k}]$ to denote the allocation to agent k with type $t^k \in T$. The expected utility of agent k with type t^k under this rule will be

---

[1] Strictly speaking, one should allow an allocation rule to be randomized. We omit this possibility.

Revenue Maximization

(under the assumption that agents announce truthfully)

$$E_{\mathbf{t}^{-k}}[v(\alpha_k[t^k,\mathbf{t}^{-k}]|t^k)] = \sum_{\mathbf{t}^{-k}} v(\alpha_k[t^k,\mathbf{t}^{-k}]|t^k)\pi(\mathbf{t}^{-k}).$$

The expected utility of agent k with type t^k announcing $\hat{t}^k$ ($\neq t^k$) under allocation rule α is

$$E_{\mathbf{t}^{-k}}[v(\alpha_k[\hat{t}^k,\mathbf{t}^{-k}]|t^k)] = \sum_{\mathbf{t}^{-k}} v(\alpha_k[\hat{t}^k,\mathbf{t}^{-k}]|t^k)\pi(\mathbf{t}^{-k}).$$

To ensure that agents will report truthfully, we impose Bayes-Nash incentive compatibility (BNIC) for each agent k with type t^k:

$$E_{\mathbf{t}^{-k}}[v(\alpha_k[t^k,\mathbf{t}^{-k}]|t^k)] - P_{t^k} \geq E_{\mathbf{t}^{-k}}[v(\alpha_k[\hat{t}^k,\mathbf{t}^{-k}]|t^k)] - P_{\hat{t}_k} \quad \forall t^k, \hat{t}_k \in T.$$

To ensure that each agent has the incentive to participate, we impose the (interim) individual rationality (IR) constraint:

$$E_{\mathbf{t}^{-k}}[v(\alpha_k[t^k,\mathbf{t}^{-k}]|t^k)] - P_{t^k} \geq 0 \;\; \forall t^k \in T.$$

If we add a dummy type t_0 which assigns utility 0 to all allocations, and set $P_{t_0} = 0$, we can fold the IR constraint into the BNIC constraint. So, from now on, T contains the dummy type t_0.

The designer's problem is to maximize expected revenue subject to BNIC and IR. To minimize notation, let us assume the optimal mechanism is symmetric. That is, if agents k and j report the same type, they face the same expected payment and allocation. Under this assumption, it is enough to focus on agent k, say. Then, per capita expected revenue is exactly the expected revenue from agent k. This is $\sum_{t^k \in T} f_{t^k} P_{t^k}$. Fix an allocation rule α and let

$$R(\alpha) = \max_{P_{t^k}} \sum_{t^k \in T} f_{t^k} P_{t^k}$$

s.t. $E_{\mathbf{t}^{-k}}[v(\alpha_k[t^k,\mathbf{t}^{-k}]|t^k)] - P_{t^k} \geq E_{\mathbf{t}^{-k}}[v(\alpha_k[\hat{t}_k,\mathbf{t}^{-k}]|t^k)] - P_{\hat{t}^k} \;\; \forall t^k, \hat{t}^k \in T.$

Call this program LP_α. If LP_α is infeasible set $R(\alpha) = -\infty$. Thus the designer's problem is to find an allocation rule α that maximizes $R(\alpha)$.

If we fix the allocation rule α, we can rewrite the BNIC constraint:

$$P_t - P_{\hat{t}} \leq E_{\mathbf{t}^{-k}}[v(\alpha_k[t,\mathbf{t}^{-k}]|t)] - E_{\mathbf{t}^{-k}}[v(\alpha_k[\hat{t},\mathbf{t}^{-k}]|t)] \;\; \forall t \neq \hat{t}.$$

The inequality system has the expected network interpretation. Introduce a node for each type (the node corresponding to the dummy type will be the source) and, to each directed edge $(\hat{t},t)$, assign a length of $E_{\mathbf{t}^{-k}}[v(\alpha_k[t,\mathbf{t}^{-k}]|t)] - E_{\mathbf{t}^{-k}}[v(\alpha_k[\hat{t},\mathbf{t}^{-k}]|t)]$. Then, P_t is upper-bounded by the length of the shortest-path from the source to vertex t. The optimal choice of P_t would be to set it equal to the length of this shortest path. For fixed α, the optimization problem reduces to determining the shortest-path tree (the union of all shortest-paths from source to all nodes) in this network. Edges on the shortest-path tree correspond to binding BNIC constraints.

Example 13 *Consider the allocation of one good among two agents with types $\{0, 1, 2\}$; here 0 is the dummy type. Let $f_1 = f_2 = 1/2$, then $\pi(1) = \pi(2) = 1/2$. Choose as the allocation rule, α, the following: assign the object to the agent with the highest nonzero type; in case of ties, randomize the allocation "50–50." The possible allocations are: agent 1 wins, agent 2 wins, agent 1 and 2 get $1/2$ of the item, the seller keeps it. An agent of type t^k who obtains the item values it at t^k; if she ties with her competitor and gets only half, she derives value $t^k/2$. If she does not receive the item (because the other agent receives it or the seller retains it), she gets value 0.*

Now to the computation of expected utility when honest. If $t^k = 1$, then

$$E_{\mathbf{t}^{-k}}[v(\alpha_k[1, \mathbf{t}^{-k}]|1)] = \sum_{\mathbf{t}^{-k}\in\{0,1,2\}} v(\alpha_k[1, \mathbf{t}^{-k}]|1)\pi(\mathbf{t}^{-k}) = \frac{1}{4},$$

and is equal to $\frac{3}{2}$ if $t^k = 2$.

Similarly, we obtain for $E_{\mathbf{t}^{-k}}[v(\alpha_k[t, \mathbf{t}^{-k}]|t^k)]$:

$t \setminus t^k$	0	1	2
0	0	0	0
1	0	$\frac{1}{4}$	$\frac{3}{4}$
2	0	$\frac{1}{2}$	$\frac{3}{2}$

So we obtain for the optimization problem

$$\max \ \tfrac{1}{2}P_1 + \tfrac{1}{2}P_2$$
$$\text{s.t.} \ P_1 - P_0 \le \tfrac{1}{4}, \ P_0 - P_1 \le 0, \ P_0 - P_2 \le 0,$$
$$P_2 - P_0 \le \tfrac{3}{2}, \ P_2 - P_1 \le 1, \ P_1 - P_2 \le -\tfrac{1}{2},$$
$$P_0 = 0.$$

The corresponding network is depicted in Figure 6.1. The dashed edges form the shortest-path tree. Reading off the shortest-path distances to nodes 1 and 2 yields $P_1 = 1/4$ and $P_2 = 5/4$. So, the designer can realize expected revenue of $3/4$ with this allocation rule.

6.1 WHAT IS A SOLUTION?

What exactly does it mean to solve an optimal auction problem? In our abstract setup, the problem of finding the optimal auction (assuming symmetry) can be written as:

$$\max_{P_{t^k},\alpha} \sum_{t^k \in T} f_{t^k} P_{t^k}$$

$$\text{s.t.} \ E_{\mathbf{t}^{-k}}[v(\alpha_k[t^k, \mathbf{t}^{-k}]|t^k)] - P_{t^k} \ge E_{\mathbf{t}^{-k}}[v(\alpha_k[\hat{t}_k, \mathbf{t}^{-k}]|t^k)] - P_{\hat{t}^k} \ \forall \, t^k, \hat{t}^k \in T.$$

$$\alpha(\mathbf{t}) \in \Gamma \ \forall \mathbf{t} \in T^n.$$

6.1 What Is a Solution?

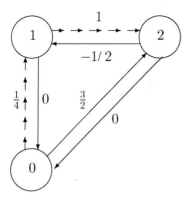

Figure 6.1 Shortest path tree.

Call this problem (OAP). If membership in Γ can be described by a set of linear inequalities and u is linear in $\alpha(\mathbf{t})$, we have a linear program and its solution is the desired optimal auction. If this is what is meant by solving an optimal auction problem, the exercise is trivial. To formulate the problem in a nontrivial way, consider the problem of finding the optimal auction when the designer knows each agent's type. Call this the **full information** problem. In this case, the BNIC constraints can be ignored. Only the IR and feasibility constraints matter. Given an allocation rule α, we maximize expected revenue by charging each agent the value of what they receive, that is,

$$E_{\mathbf{t}^{-k}}[v(\alpha_k[t^k, \mathbf{t}^{-k}]|t^k)] = P_{t^k}.$$

Thus, in the full information case, the problem of finding the optimal auction reduces to solving

$$\max_{\alpha} \sum_{t^k \in T} f_{t^k} E_{\mathbf{t}^{-k}}[v(\alpha_k[t^k, \mathbf{t}^{-k}]|t^k)]$$

$$\alpha(\mathbf{t}) \in \Gamma \;\; \forall \mathbf{t} \in T^n.$$

Notice that the full information problem involves just the α variables, a potentially simpler problem. We will say that problem (OAP) is solved if it can be reduced to an optimization problem involving just the α variables. The goal in the following sections is to identify cases when such a reduction can be effectuated.[2]

---

[2] One might ask for more. For example, problem (OAP) can be solved by some "simple" algorithm. This is too much. If the underlying full information problem is extremely complex to solve, there is no reason to suppose that the corresponding instance of (OAP) would be easier.

6.2 ONE-DIMENSIONAL TYPES

The most well-understood case is when the type space, T, is one-dimensional. Here we take it to be discrete, that is, $T = \{1, 2, \ldots, m\}$.[3] Let Γ, the space of feasible allocations, be a subset of Euclidean space.[4] Let **a** be an allocation rule. We restrict attention to allocation rules that are anonymous, that is, independent of the names of agents. This restriction is without loss when each agent's type is a draw from the same distribution and there is appropriate symmetry in Γ.

Denote by **t** a profile of types and $\mathbf{a}_i(t)$ the allocation to a type $i \in T$ in profile **t**. It will be convenient to restrict attention to one agent – say, agent 1. Write $\mathcal{A}_i$ to be the **expected allocation** (also called the **interim allocation**) that agent 1 with type i receives, that is,

$$\mathcal{A}_i = \sum_{\mathbf{t}^{-1} \in T^{n-1}} \mathbf{a}_i[i, \mathbf{t}^{-1}] \pi(\mathbf{t}^{-1}). \tag{6.1}$$

Example 14 *Consider the allocation of a single good, where the type of an agent is their marginal value for the good. Suppose the allocation rule assigns the object with equal probability to the agents with the highest type. This is called the **efficient** allocation rule. In this case, we interpret $\mathbf{a}_i(\mathbf{t})$ to be the probability that an agent with type i in profile $\mathbf{t}$ receives the good. If the good were divisible, we could interpret $\mathbf{a}_i(\mathbf{t})$ to be the fraction of the good that an agent with type i in profile $\mathbf{t}$ receives.*

If $n_j(\mathbf{t})$ is the number of agents with type j in profile $\mathbf{t}$ and i is the largest type in the profile, then $\mathbf{a}_i(\mathbf{t}) = \frac{1}{n_i(\mathbf{t})}$.

To determine $\mathcal{A}_i$ for any i, observe that in any profile $\mathbf{t}$ where $n_i(\mathbf{t}) > 0$, an agent with type i will receive the good with positive probability provided $n_j(\mathbf{t}) = 0$ for all $j > i$. Further, if $n_j(\mathbf{t}) = 0$ for all $j > i$, then, $\mathbf{a}_i(\mathbf{t}) = \frac{1}{n_i(\mathbf{t})}$. Hence, given agent 1 is of type i, the probability of drawing a profile where there are $k - 1$ other agents with type i and $(n - 1) - (k - 1)$ agents with type $i - 1$ or smaller is $\binom{n-1}{k-1} f_i^{k-1} F(i-1)^{n-k}$. Therefore,

$$\mathcal{A}_i = \sum_{k=1}^{n} \frac{1}{k} \binom{n-1}{k-1} f_i^{k-1} F(i-1)^{n-k} = \left(\frac{1}{nf_i}\right) \sum_{k=1}^{n} \binom{n}{k} f_i^k F(i-1)^{n-k}$$

$$= \left(\frac{1}{nf_i}\right) \sum_{k=0}^{n} \binom{n}{k} f_i^k F(i-1)^{n-k} - \left(\frac{1}{nf_i}\right) F(i-1)^n = \frac{F(i)^n - F(i-1)^n}{nf_i}.$$

We now verify that this choice of $\mathcal{A}_i$ is monotone. Observe that

$$F(i)^n - F(i-1)^n = [F(i) - F(i-1)] \sum_{j=0}^{n-1} F(i)^{n-j-1} F(i-1)^j$$

$$= f_i \sum_{j=0}^{n-1} F(i)^{n-j-1} F(i-1)^j.$$

---

[3] This case is also examined in Rochet and Stole (2003), Lovejoy (2006), and Elkind (2007).
[4] One could generalize a little further by allowing Γ to be a lattice. However, the additional notation is not worth the burden.

6.2 One-Dimensional Types

Hence,

$$A_i = \left(\frac{1}{n}\right) \sum_{j=0}^{n-1} F(i)^{n-j-1} F(i-1)^j,$$

which is clearly monotone in i because $F(i)$ is monotone in i.

In an abuse of notation, we write $v(\mathcal{A}_i|i)$ to mean the expected value that agent 1 of type i receives with a lottery over allocations with expected value $\mathcal{A}_i$. Specifically,

$$v(\mathcal{A}_i|i) = \sum_{\mathbf{t}^{-1} \in T^{n-1}} v(\mathbf{a}_i[i, \mathbf{t}^{-1}]|i) \pi(\mathbf{t}^{-1}).$$

Following Myerson (1981), one eliminates the P_j variables by showing that they can be written as a linear function of the $\mathcal{A}_i$'s (essentially revenue equivalence). This leaves one with an optimization problem in just the $\mathcal{A}_i$'s variables. At this stage there are two approaches one can take. Myerson (1981) uses equation (6.1) to convert back into the **a** variables and work directly with those. The second approach, emphasized here, uses a polyhedral characterization of the $\mathcal{A}$ variables that satisfy equation (6.1). This allows one to solve the problem using just the $\mathcal{A}_i$ variables. An example is provided later in this chapter to illustrate the advantages of the second method.

Now, BNIC requires

$$v(\mathcal{A}_i|i) - P_i \geq v(\mathcal{A}_j|i) - P_j.$$

Associate a network in the usual way with these BNIC constraints. Introduce a vertex i for each type i. Between every ordered pair of vertices (j, i) insert a directed edge of length $v(\mathcal{A}_i|i) - v(\mathcal{A}_j|i)$. The allocation rule will be IBN if this network has no negative-length cycle. If this network has no negative-length cycles, we can choose the P_i's to be the length of the shortest path from an arbitrarily chosen root vertex, r, to vertex i in the shortest-path tree rooted at r. To ensure that the IR constraint holds, choose the root r to, be the vertex corresponding to the dummy type t_0.

Theorem 6.2.1 *Suppose that v is nondecreasing and satisfies strictly increasing differences. An allocation rule **a** is IBN iff. the corresponding expected allocations are monotonic. That is, if $r \leq s$, then $\mathcal{A}_r \leq \mathcal{A}_s$.*

Proof. Same as Theorem 4.2.5. ∎

If $i \geq j$, we refer to

$$v(\mathcal{A}_i|i) - P_i \geq v(\mathcal{A}_j|i) - P_j$$

as a downward BNIC constraint. If $i < j$, it is called an upward BNIC constraint. Observe that both upward *and* downward BNIC constraints are needed to establish that $\mathcal{A}$ is monotone.

Next we show that "adjacent" BNIC constraints suffice.

116 Revenue Maximization

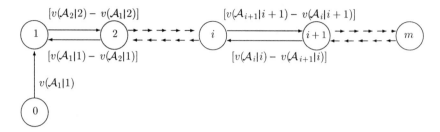

Figure 6.2 Network of BNIC constraints

Theorem 6.2.2 *Suppose that v satisfies strictly increasing differences. All BNIC constraints are implied by the following:*

$$v(\mathcal{A}_i|i) - P_i \geq v(\mathcal{A}_{i-1}|i) - P_{i-1} \quad \forall i = 1, \ldots, m \qquad (\text{BNIC}_i^d)$$

$$v(\mathcal{A}_i|i) - P_i \geq v(\mathcal{A}_{i+1}|i) - P_{i+1} \quad \forall i = 0, \ldots, m-1 \qquad (\text{BNIC}_i^u)$$

Proof. Combining BNIC_i^d and BNIC_{i-1}^u along with the strictly increasing differences property of v is enough to conclude that $\mathcal{A}_i$ is monotone (see proof of Theorem 4.2.5). Hence, the length of the edge from i to $i+2$ must be at least as large as the length of $(i, i+1)$ plus the length of $(i+1, i+2)$.

In view of the above, the network associated with the BNIC constraints can be depicted as shown in Figure 6.2. ∎

Suppose that $\mathcal{A}$ is IBN. Then, the network of Figure 6.2 has no negative-length cycles. In this case, it is easy to see that the shortest-path tree rooted at dummy vertex '0' must be $0 \to 1 \to 2 \to \cdots \to m$. Algebraically, set $\mathcal{A}_0 = 0$ and $P_0 = 0$ for the dummy type and

$$P_i = \sum_{r=1}^{i} [v(\mathcal{A}_r|r) - v(\mathcal{A}_{r-1}|r)]. \qquad (6.2)$$

Notice that $P_i - P_{i-1} = v(\mathcal{A}_i|i) - v(\mathcal{A}_{i-1}|i)$ for the above expected payment schedule, hence all downward BNIC constraints are satisfied and bind, that is,

$$v(\mathcal{A}_i|i) - P_i = v(\mathcal{A}_{i-1}|i) - P_{i-1} \quad \forall i = 1, \ldots, m.$$

It is easy to see that the upward BNIC constraints all hold.

We now summarize our conclusions.

Theorem 6.2.3 *For any monotonic $\mathcal{A}$, there exists an expected payment schedule $\{P_i\}_{i=0}^{m}$, such that all the adjacent BNIC constraints are satisfied.*

Proof. Set $\mathcal{A}_0 = 0$ and $P_0 = 0$ for the dummy type. Then set

$$P_i = \sum_{r=1}^{i} [v(\mathcal{A}_r|r) - v(\mathcal{A}_{r-1}|r)].$$

6.2 One-Dimensional Types

Notice that $P_i - P_{i-1} = v(\mathcal{A}_i|i) - v(\mathcal{A}_{i-1}|i)$ for the above expected payment schedule $\{P_i\}_{i=0}^m$, hence all (BNIC^d) are satisfied and bind, that is,

$$v(\mathcal{A}_i|i) - P_i = v(\mathcal{A}_{i-1}|i) - P_{i-1} \quad \forall i = 1, \ldots, m.$$

Hence, all (BNIC^u) are satisfied, and thus, by Theorem 6.2.2, the allocation rule **a** is incentive compatible. ∎

6.2.1 A Formulation

Because P_i is the expected payment received from agent 1 when she claims type i and the allocation rule is anonymous, the expected payment *per agent* will be $\sum_{i=0}^m f_i P_i$. Hence, the problem of finding the revenue-maximizing mechanism can be formulated as:

$$Z = \max_{\{\mathbf{a}\}} n \sum_{i=0}^m f_i P_i \quad (OPT1)$$

s.t. $v(\mathcal{A}_i|i) - P_i \geq v(\mathcal{A}_{i-1}|i) - P_{i-1} \quad \forall i = 1, \ldots, m$ (BNIC^d_i)

$v(\mathcal{A}_i|i) - P_i \geq v(\mathcal{A}_{i+1}|i) - P_{i+1} \quad \forall i = 0, \ldots, m-1$ (BNIC^u_i)

$0 = \mathcal{A}_0 \leq \mathcal{A}_1 \leq \cdots \leq \mathcal{A}_i \leq \cdots \leq \mathcal{A}_m$ (M)

$P_0 = 0$ (Z)

$\mathcal{A}_i = \sum_{\mathbf{t}^{-1} \in T^{n-1}} \mathbf{a}_i[i, \mathbf{t}^{-1}] \pi(\mathbf{t}^{-1})$ (A)

$\mathbf{a}(\mathbf{t}) \in \Gamma \quad \forall \mathbf{t} \in T^n.$ (C)

An upper bound on each P_i is the length of the shortest path from the dummy node "0" to vertex i in the network of Figure 6.2. This is stated below.

Lemma 6.2.4 *All downward constraints (BNIC^d_i) bind in a solution to the (OPT1) problem.*

The essence of the previous result is that once the allocation rule is chosen, equation (6.2) pins down the payments necessary to ensure both incentive compatibility and maximum expected revenue. Our problem reduces to finding the optimal allocation rule.

Given equation (6.2),

$$\sum_{i=1}^m f_i P_i = \sum_{i=1}^m f_i \sum_{r=1}^i [v(\mathcal{A}_r|r) - v(\mathcal{A}_{r-1}|r)].$$

Write $F(i) = \sum_{r=1}^i f_r$ and $F(m) = 1$. Then,

$$\sum_{i=1}^m f_i P_i = \sum_{i=1}^m \{f_i v(\mathcal{A}_i|i) + (1 - F(i))[v(\mathcal{A}_i|i) - v(\mathcal{A}_i|i+1)]\}.$$

We interpret $v(\mathcal{A}_i|m+1)$ to be zero.
Let
$$\mu(\mathcal{A}_i) = v(\mathcal{A}_i|i) - \left(\frac{1-F(i)}{f_i}\right)[v(\mathcal{A}_i|i+1) - v(\mathcal{A}_i|i)].$$
Myerson (1981) calls $\mu(\mathcal{A}_i)$ a **virtual valuation**.

Problem (*OPT*1) becomes

$$Z = \max_{\{\mathbf{a}\}} n \sum_{i=1}^{m} f_i \mu(\mathcal{A}_i) \qquad (OPT2)$$

s.t. $0 \leq \mathcal{A}_1 \leq \cdots \leq \mathcal{A}_i \leq \cdots \leq \mathcal{A}_m$

$$\mathcal{A}_i = \sum_{\mathbf{t}^{-i} \in T^{n-1}} \mathbf{a}_i[i, \mathbf{t}^{-i}] \pi(\mathbf{t}^{-i})$$

$\mathbf{a}(\mathbf{t}) \in \Gamma \quad \forall \mathbf{t} \in T^n.$

In the absence of any structure on $\mu_i(\mathcal{A}_i)$, there is little more to be said. It is usual to suppose the underlying environment is such that $\mu_i(\mathcal{A}_i)$ is linear in $\mathcal{A}_i$. We examine one such instance below.

Before continuing, it is important to emphasize that the analysis needed to obtain (*OPT*2) did not depend heavily on the structure of Γ. Indeed, it suffices that the set of possible allocations form a lattice and the valuations have a suitable monotonicity and increasing differences property on that lattice.

6.2.2 Optimal Mechanism for Sale of a Single Object

We focus now on a specific instance, that of deriving the revenue-maximizing auction for the sale of a single object to buyers with constant marginal values. We derive the revenue-maximizing mechanism in two ways. The first follows Myerson (1981).[5] In this environment, the type of a buyer is their marginal value for the good, and $\mathbf{a}_i(\mathbf{t})$ is the probability that agent 1 receives the good when his type is i and the reported profile types is $\mathbf{t}$. In this case, $\mathcal{A}_i$ is the expected probability that an agent with type i receives the object.

The constant marginal value assumption means that $v(\mathcal{A}_i|i) = i\mathcal{A}_i$. This ensures that $\mu_i(\mathcal{A}_i)$ is linear in $\mathcal{A}_i$. Specifically,

$$\mu(\mathcal{A}_i) = \mathcal{A}_i \left(i - \frac{1-F(i)}{f_i}\right).$$

See Figueroa and Skreta (2009) for a discussion of what can happen when $\mu_i(\mathcal{A}_i)$ is not linear in $\mathcal{A}_i$.

It is also usual to assume $\frac{1-F(i)}{f_i}$, is nonincreasing in i. This is called the **monotone hazard condition** and many natural distributions satisfy it. This is an assumption for arithmetical convenience only and we relax it later.

---

[5] Other papers that consider the same problem are Harris and Raviv (1981) and Riley and Samuelson (1981).

6.2 One-Dimensional Types

Because the expression $i - \frac{1-F(i)}{f_i}$ will be ubiquitous, set $v(i) = i - \frac{1-F(i)}{f_i}$. Under the monotone hazard condition, $v(i)$ is nondecreasing in i. Let $n_i(\mathbf{t})$ be the number of agents with type i in the profile $\mathbf{t}$. The problem of finding the revenue-maximizing auction can be formulated as:

$$Z_1 = \max_{\{\mathbf{a}\}} n \sum_{i=1}^{m} f_i \mathcal{A}_i v(i) \qquad (OPT3)$$

$$\text{s.t. } 0 \leq \mathcal{A}_1 \leq \cdots \leq \mathcal{A}_i \leq \cdots \leq \mathcal{A}_m$$

$$\mathcal{A}_i = \sum_{\mathbf{t}^{-1} \in T^{n-1}} \mathbf{a}_i[i, \mathbf{t}^{-1}] \pi(\mathbf{t}^{-1}) \; \forall i$$

$$\sum_{i=1}^{m} n_i(\mathbf{t}) \mathbf{a}_i(\mathbf{t}) \leq 1 \; \forall \mathbf{t} \in T^n.$$

Program (*OPT3*) can be rewritten to read:

$$\max_{\{\mathbf{a}\}} n \sum_{i=1}^{m} \sum_{\mathbf{t}^{-1} \in T^{n-1}} f_i \mathbf{a}_i[i, \mathbf{t}^{-1}] v(i) \pi(\mathbf{t}^{-1}) \qquad (OPT4)$$

$$\text{s.t. } 0 \leq \sum_{\mathbf{t}^{-1} \in T^{n-1}} \mathbf{a}_1[1, \mathbf{t}^{-1}] \pi(\mathbf{t}^{-1}) \leq \cdots \leq \sum_{\mathbf{t}^{-1} \in T^{n-1}} \mathbf{a}_m[m, \mathbf{t}^{-1}] \pi(\mathbf{t}^{-1}) \leq 1$$

$$\sum_{i=1}^{m} n_i(\mathbf{t}) \mathbf{a}_i(\mathbf{t}) \leq 1 \; \forall \mathbf{t} \in T^n.$$

If the monotonicity constraints are ignored, then problem (*OPT4*) can be decomposed into $|T^n|$ subproblems, one for each profile $\mathbf{t}$ of types:

$$\max_{\{\mathbf{a}\}} \sum_i n_i(\mathbf{t}) v(i) \mathbf{a}_i(\mathbf{t}) \qquad (OPT5)$$

$$\text{s.t. } \sum_{i=1}^{m} n_i(\mathbf{t}) \mathbf{a}_i(\mathbf{t}) \leq 1$$

Problem (*OPT5*) is an instance of a (fractional) knapsack problem. Its solution is well known: select the variable with largest "bang for the buck" ratio (provided it is positive) and make it as large as possible. Formally, select any type

$$r \in \arg\max\{\frac{n_i(\mathbf{t}) v(i)}{n_i(\mathbf{t})} : n_i(\mathbf{t}) > 0, v(i) \geq 0\}$$

and set $a_r(\mathbf{t}) = 1/n_r(\mathbf{t})$. The monotone hazard condition ensures that the largest type is always selected provided its virtual value is nonnegative. Thus, the solution to the program is monotonic, that is, $\mathbf{a}_{i+1}(\mathbf{t}) \geq \mathbf{a}_i(\mathbf{t})$ for all i and profiles t. It follows from this that the ignored monotonicity constraints on expected allocations are satisfied.

This analysis identifies the optimal allocation rule. One must still compute the relevant payments to obtain a complete description of the entire mechanism. From equation (6.2), we know for the assumed functional form of v, that

$$P_i = i\mathcal{A}_i - \sum_{j=0}^{i-1} \mathcal{A}_j.$$

Thus, to compute expected payments, we must compute expected allocations implied by the allocation rule.[6] To do so, let i^* be the smallest type such that $v(i^*) \geq 0$. Since $\mathbf{a}_i(\mathbf{t}) = 0$ for all $i < i^*$, it follows that $\mathcal{A}_i = 0$ for all $i < i^*$. To determine $\mathcal{A}_i$ for any $i \geq i^*$, observe that in any profile $\mathbf{t}$ where $n_i(\mathbf{t}) > 0$, an agent with type i will receive the good with positive probability provided $n_j(\mathbf{t}) = 0$ for all $j > i$. Further, if $n_j(\mathbf{t}) = 0$ for all $j > i$, then $\mathbf{a}_i(\mathbf{t}) = \frac{1}{n_i(\mathbf{t})}$. Hence, given agent 1 is of type i, the probability of drawing a profile where there are $k - 1$ other agents with type i and $(n - 1) - (k - 1)$ agents with type $i - 1$ or smaller is $\binom{n-1}{k-1} f_i^{k-1} F(i-1)^{n-k}$. Therefore,

$$\mathcal{A}_i = \sum_{k=1}^{n} \frac{1}{k} \binom{n-1}{k-1} f_i^{k-1} F(i-1)^{n-k} = \frac{F(i)^n - F(i-1)^n}{nf_i}.$$

Hence, for all $i \geq i^* + 1$,

$$P_i = i\left[\frac{F(i)^n - F(i-1)^n}{nf_i}\right] - \sum_{j=i^*}^{i-1} \frac{F(j)^n - F(j-1)^n}{nf_j} \qquad (6.3)$$

and $P_{i^*} = i^*\left[\frac{F(i^*)^n - F(i^*-1)^n}{nf_{i^*}}\right]$.

The mechanism can be described thusly: Each agent reports their type. An agent who reports type i is charged according to equation (6.3), whether she receives the good or not. If the virtual value of the reported type is negative, the agent receives nothing (but also pays nothing, according to the formula). If the agent's virtual value is positive but not the largest reported, she receives nothing (note that she still makes a payment). If the agent's virtual value is among the highest reported, she shares the object with equal probability among the other agents with the same virtual value.

Dominant Strategy

In fact, as the reader may verify, the allocation rule associated with the optimal mechanism satisfies the stronger condition of being dominant strategy incentive compatible. This is because holding the types of all agents except agent 1 fixed, the probability of agent 1 receiving the good is nondecreasing in his type.

---

[6] This will mimic exactly example 14.

6.2 One-Dimensional Types

Except for the proviso that agents with types below i^* receive nothing, it is identical with the efficient allocation rule discussed in example 14. Further, there exists a payment rule that implements this allocation. The winning agent is charged the larger of the second-highest type and P_{i^*}. Here P_{i^*} is given by equation (6.3) and is called the reserve price.

This payment rule satisfies a stronger version of individual rationality: ex-post individual rationality. That is, the ex-post surplus of every agent from participating in the mechanism at every profile is nonnegative. With these observations in hand, it is possible to derive, in the same way, the revenue-maximizing, ex-post individually rational, and dominant strategy mechanism for selling a single object. By direct computation, one may verify that this mechanism generates the same expected revenue as the optimal Bayesian incentive compatible mechanism. Manelli and Vincent (2008) demonstrate that this equivalence between Bayesian and dominant strategy incentive compatible mechanisms holds more widely, as we discuss later.

6.2.3 Polyhedral Approach

An alternative approach to the solution of problem (*OPT*3) with other applications is based on a polyhedral characterization of the set of expected allocations due to Border (1991).[7] Call a set of $\mathcal{A}_i$'s feasible if there exists an allocation rule **a** such that

$$\mathcal{A}_i = \sum_{\mathbf{t}^{-1} \in T^{n-1}} \mathbf{a}_i[i, \mathbf{t}^{-1}] \pi(\mathbf{t}^{-1}) \ \forall i.$$

Theorem 6.2.5 (Border's Theorem) *The expected allocation $\mathcal{A}_i$ is feasible iff.*

$$n \sum_{i \in S} f_i \mathcal{A}_i \leq 1 - (\sum_{i \notin S} f_i)^n \ \forall S \subseteq \{1, 2, \ldots, m\}. \tag{6.4}$$

Proof. We first prove necessity. Fix a set S of types. Let E be the set of n agent profiles where at least one agent has a type in S. Now,

$$n \sum_{i \in S} f_i \mathcal{A}_i = n \sum_{i \in S} f_i [\sum_{\mathbf{t}^{-1} \in T^{n-1}} \mathbf{a}_i[i, \mathbf{t}^{-1}] \pi(\mathbf{t}^{-1})]$$

$$= \sum_{j=1}^{n} \sum_{i \in S} [\sum_{(i,\mathbf{t}^{-j}) \in T^n} \mathbf{a}_i[i, \mathbf{t}^{-j}] f_i \pi(\mathbf{t}^{-j})] = \sum_{\mathbf{t} \in E} \pi(\mathbf{t}) \sum_{i \in S} n_i(\mathbf{t}) \mathbf{a}_i(\mathbf{t})$$

$$\leq \sum_{\mathbf{t} \in E} \pi(\mathbf{t}) = \pi(\mathbf{t} \in E) = 1 - (\sum_{i \notin S} f_i)^n.$$

---

[7] Border originally established this result when types were continuous and the allocation rules were anonymous. A subsequent paper, Border (2007), deals with the discrete type case and relaxes the anonymity condition. Mierendorff (2009) relaxes the anonymity condition for continuous types.

The last inequality follows from the fact that the set of allocation rules, **a**, satisfies

$$\sum_i n_i(\mathbf{t})\mathbf{a}_i(\mathbf{t}) \leq 1 \quad \forall \mathbf{t} \in T^n.$$

This implies that the set of allocation rules forms a closed convex set. Hence, the set of feasible expected allocations is also a closed convex set. Therefore, if $\mathcal{A}'$ satisfies (6.4), but is not a feasible expected allocation, there is a hyperplane that strictly separates $\mathcal{A}'$ from the set of feasible expected allocations. Specifically, there are $(c_1, c_2, \ldots, c_m)$, all distinct and nonzero, such that

$$\sum_{i=1}^{m} c_i \mathcal{A}_i' > \sum_{i=1}^{m} c_i \mathcal{A}_i \tag{6.5}$$

for all feasible expected allocations $\mathcal{A}$. Because $\mathcal{A} = 0$ is a feasible expected allocation, it follows from (6.5) that at least one $c_i > 0$. It is convenient to set $h_i = \frac{c_i}{f_i}$, and one can choose the c_i's to satisfy inequality (6.5) so that the h_i are all distinct. Rewrite inequality (6.5) as

$$\sum_{i=1}^{m} h_i f_i \mathcal{A}_i' > \sum_{i=1}^{m} h_i f_i \mathcal{A}_i. \tag{6.6}$$

Let $H^+ = \{i : h_i > 0\}$ and $H^- = \{i : h_i < 0\}$. Consider the allocation rule $\mathbf{a}^*$ that for each profile assigns the good (with equal probability) to agents whose type $i \in H^+$ has the highest h_i, provided one exists. If S_r is a set of types that correspond to the $r \leq |H^+|$ largest h_i values, then a straightforward calculation establishes that

$$n \sum_{i \in S_r} f_i \mathcal{A}_i^* = 1 - (\sum_{j \notin S_r} f_j)^n. \tag{6.7}$$

Because $\mathcal{A}^*$ is feasible and $\mathcal{A}'$ is not, it follows from (6.6) that

$$\sum_{i=1}^{m} h_i f_i \mathcal{A}_i' > \sum_{i=1}^{m} h_i f_i \mathcal{A}_i^* \Rightarrow \sum_{i \in H^+} h_i f_i \mathcal{A}_i' > \sum_{i \in H^+} h_i f_i \mathcal{A}_i^*, \tag{6.8}$$

where the last inequality follows from the fact that $h_i < 0$ and $\mathcal{A}_i^* = 0$ for all $i \in H^-$.

Let $s = |H^+|$, and for each $1 \leq r \leq s$ let q_r be the rth smallest h_i value in H^+. For convenience, set $q_{s+1} = 0$. Notice that $q_1 > q_2 > \cdots > q_s > q_{s+1} = 0$. Then

$$\sum_{i \in H^+} h_i f_i \mathcal{A}_i' = \sum_{r=1}^{s} (q_r - q_{r+1}) [\sum_{i \in S_r} f_i \mathcal{A}_i']$$

and

$$\sum_{i \in H^+} h_i f_i \mathcal{A}_i^* = \sum_{r=1}^{s} (q_r - q_{r+1}) [\sum_{i \in S_r} f_i \mathcal{A}_i^*].$$

Hence, inequality (6.8) implies that

$$\sum_{r=1}^{s}(q_r - q_{r+1})[\sum_{i\in S_r} f_i \mathcal{A}'_i] > \sum_{r=1}^{s}(q_r - q_{r+1})[\sum_{i\in S_r} f_i \mathcal{A}^*_i].$$

Therefore, there is an index r such that

$$\sum_{i\in S_r} f_i \mathcal{A}'_i > \sum_{i\in S_r} f_i \mathcal{A}^*_i = \frac{1 - (\sum_{j\notin S_r} f_j)^n}{n},$$

where the last equation follows from (6.7). However, this contradicts the assumption that $\mathcal{A}'$ satisfies (6.4). ∎

Three remarks are in order. First, the theorem makes crucial use of the independence assumption on distribution of types. Second, no use is made of the fact that types are one-dimensional, that is, Border's Theorem holds even for multidimensional types. Third, if the feasible expected allocations are monotone and types one-dimensional, then the following is a straight forward consequence based on the last part of the proof of Border's Theorem.

COROLLARY 6.2.6 *An expected allocation $\mathcal{A}_i$ that is monotone in type is feasible iff.*

$$n\sum_{i\geq r} f_i \mathcal{A}_i \leq 1 - (\sum_{i<r} f_i)^n = 1 - F(r-1)^n \ \forall r \in \{1, 2, \ldots, m\}. \quad (6.9)$$

Using (6.9), we can reformulate the problem of finding the revenue-maximizing mechanism for the sale of a single object as follows:

$$Z_1 = \max\ n \sum_{i=1}^{m} f_i \mathcal{A}_i v(i) \qquad (OPT6)$$

$$\text{s.t.} \sum_{i\geq r} f_i \mathcal{A}_i \leq \frac{1 - F(r-1)^n}{n} \quad \forall r \in \{1, 2, \ldots, m\}$$

$$0 \leq \mathcal{A}_1 \leq \cdots \leq \mathcal{A}_i \leq \cdots \leq \mathcal{A}_m$$

Even though (*OPT6*) appears forbidding, appearances can be deceiving. We discuss the solution to a special class of linear programs of which (*OPT6*) is an instance. The linear program has the form:

$$\max \sum_{j=1}^{m} c_j x_j \quad (P_\Delta)$$

$$\text{s.t.} \sum_{j\geq r} a_j x_j \leq b_r \quad \forall r \in \{1, \ldots, m\}$$

$$x_j \geq 0 \quad \forall j.$$

There are three conditions we impose on problem P_Δ:

(1) $b_1 \geq b_2 \geq \cdots \geq b_m$;
(2) $a_j > 0$ for all j;
(3) $c_1/a_1 \leq c_2/a_2 \leq \ldots c_m/a_m$.

Theorem 6.2.7 *There is an optimal solution x^* to P_Δ such that*

(1) $c_j < 0 \Rightarrow x_j^* = 0$;
(2) $\sum_{j \geq r} a_j x_j^* = b_r$ for all $r \geq i^*$ where i^* is the smallest index such that $c_{i^*} > 0$;
(3) $x_m^* = \frac{b_m}{a_m}$ and $x_i^* = \frac{b_i - b_{i+1}}{a_i}$ for all $i^* \leq i \leq m - 1$.

Proof. The first part of the theorem is obvious. To prove the second part, suppose an optimal solution x^* and let r be the largest index such that $\sum_{j \geq r} a_j x_j^* < b_r$. Construct a new solution, x', from x^* by increasing x_r^* by ϵ/a_r and decreasing x_{r-1}^* by ϵ/a_{r-1}. Choose ϵ so that $\sum_{j \geq r} a_j x_j' = b_r$. Observe that x' is feasible. In addition x' binds at all constraints that x^* does. Furthermore, $cx' - cx^* = \frac{c_r}{a_r}\epsilon - \frac{c_{r-1}}{a_{r-1}}\epsilon \geq 0$. Thus we have a new solution, with an objective function value no smaller than before, that binds at more constraints. Item 3 of the theorem follows from the fact that the constraint matrix is upper triangular. ∎

Now return to program (*OPT6*). Ignoring the monotonicity constraints on the $\mathcal{A}_i$'s and assuming the monotone hazard condition, problem (*OPT6*) is an instance of program (P_Δ). Hence, at optimality, $\mathcal{A}_i = 0$ whenever $v(i) < 0$. Let i^* be the lowest type such that $v(i^*) \geq 0$.

For all $r \geq i^*$, we have $\sum_{i \geq r} f_i \mathcal{A}_i = \frac{1 - F(r-1)^n}{n}$, in particular at optimality $\mathcal{A}_m = \frac{1 - F(m-1)^n}{n f_m}$ and

$$\mathcal{A}_r = \frac{1 - F(r-1)^n}{n f_r} - \frac{1 - F(r)^n}{n f_r} = \frac{F(r)^n - F(r-1)^n}{n f_r}$$

for all $i^* \leq r \leq m - 1$. Note the similarity to example 14. Monotonicity follows as in example 14.

Because the optimal solution to the relaxation of (*OPT6*) is feasible (because it is monotone) for (*OPT6*), it must be optimal for problem (*OPT6*). Note that the solution via the inequalities in Border's Theorem only fix the value of the $\mathcal{A}_i$'s. They do not yield the actual allocation rule **a**. By staring at the form of the optimal $\mathcal{A}_i$'s, one may be able to deduce the underlying allocation rule, but this is unsatisfying. In a subsequent section, we show how to systematically deduce the underlying allocation rule when the expected allocations are extreme points of the inequalities in Border's Theorem.

Polymatroids

Problem (*OPT6*) is an instance of the problem of maximizing a linear function over a polymatroid. Let G be a nonnegative, real valued function defined on the

6.2 One-Dimensional Types

subsets of $\{1, 2, \ldots, m\}$. G is **submodular** if for all $S \subset T \subseteq \{1, 2, \ldots, m\}$ and $j \notin T$:

$$G(S \cup j) - G(S) \geq G(T \cup j) - G(T).$$

A polymatroid optimization problem is a linear program of the following kind:

$$\max \sum_{j=1}^{m} c_j x_j$$

$$\text{s.t.} \sum_{j \in S} x_j \leq G(S) \quad \forall S \subseteq \{1, 2, \ldots, m\}$$

$$x_j \geq 0 \quad \forall j \in \{1, 2, \ldots, m\}$$

The feasible region of this linear program is called a **polymatroid**.

For our purposes, define

$$G(S) = \frac{1 - (\sum_{i \notin S} f_i)^n}{n} \quad \forall S \subseteq \{1, 2, \ldots, m\}.$$

It is straightforward to verify that G is a nondecreasing submodular function. Set $x_i = n f_i \mathcal{A}_i$ for all i. Then, using Theorem 6.4, we can reformulate the problem of finding the revenue-maximizing mechanism for the sale of a single object as

$$\max n \sum_{i=1}^{m} v(i) x_i$$

$$\text{s.t.} \sum_{i \in S} x_i \leq G(S) \quad \forall S \subseteq \{1, 2, \ldots, m\}$$

$$x_i \geq 0, \quad \forall i \in \{1, 2, \ldots, m\}.$$

It is well known that polymatroid optimization problems can be solved by a greedy procedure (see, for example, III.3 of Nemhauser and Wolsey 1988).[8] Choose the index i with the largest coefficient in the objective function and increase x_i to its maximum extent possible. In our case, this would be the type with largest virtual value. Under the monotone hazard rate assumption, this would be variable x_m. From the inequalities that define the polymatroid, we see that the most x_m could be set to is $G(\{n\})$. Fix the value of x_m at $G(\{m\})$ and "move" it over to the right-hand side and repeat. Under such a procedure, x_{m-1} would be set to $G(\{m, m-1\}) - G(\{m\})$. Submodularity of G implies that $x_{m-1} = G(\{m, m-1\}) - G(\{m\}) \leq G(\{m\}) = x_m$. In general, $x_j = G(\{m, m-1, \ldots, j\}) - G(\{m, m-1, \ldots, j+1\})$ for all j. Submodularity implies that $x_m \geq x_{m-1} \geq \cdots \geq x_1$. Using the argument at the tail end of example 14, it follows that the $\mathcal{A}_i$ will be monotone.

---

[8] In fact, the proof of Theorem 6.4 can be reinterpreted as a proof of this fact.

In fact, the greedy algorithm applies in the more general case where $\mu(\mathcal{A}_i)$ is concave in $\mathcal{A}_i$ rather than linear (see, for example, Groenvelt 1991).

6.2.4 Ironing and Extreme Points

In this section, we discuss how to solve problem (*OPT6*) when the monotone hazard condition fails.[9] A by-product of this analysis will be a characterization of the extreme points of (*OPT6*). Recall that:

$$Z_1 = \max \sum_{i=1}^{m} f_i \mathcal{A}_i v(i) \qquad (OPT6)$$

$$\text{s.t.} \sum_{i \geq r} f_i \mathcal{A}_i \leq \frac{1 - F(r-1)^n}{n} \quad \forall r \in \{1, 2, \ldots, m\}$$

$$0 \leq \mathcal{A}_1 \leq \cdots \leq \mathcal{A}_i \leq \cdots \leq \mathcal{A}_m.$$

For convenience, we have omitted to scale the objective function by n, the number of agents. In the absence of the monotone hazard condition, it is possible that at optimality, some of the monotonicity constraints, $\mathcal{A}_i \leq \mathcal{A}_{i+1}$, will bind. When this happens, it is called **pooling**.

Consider the following Lagrangean relaxation of (*OPT6*):

$$L_1(\lambda) = \max \sum_{i=1}^{m} f_i \mathcal{A}_i v(i) + \sum_{i=1}^{m-1} \lambda_i [\mathcal{A}_{i+1} - \mathcal{A}_i]$$

$$\text{s.t.} \sum_{i \geq r} f_i \mathcal{A}_i \leq \frac{1 - F(r-1)^n}{n} \quad \forall r \in \{1, 2, \ldots, m\} \qquad (6.10)$$

$$\mathcal{A}_i \geq 0 \,\, \forall i = 1, \ldots, m.$$

By the duality theorem, $Z_1 = \min_{\lambda \geq 0} L_1(\lambda)$. The objective function of the program that defines $L_1(\lambda)$ can be written as

$$\sum_{i=1}^{m} f_i \mathcal{A}_i \left(v(i) - \frac{\lambda_i}{f_i} + \frac{\lambda_{i-1}}{f_i} \right)$$

where we interpret λ_m and λ_0 to be zero. For economy of notation, set

$$h_i(\lambda) = v(i) - \frac{\lambda_i}{f_i} + \frac{\lambda_{i-1}}{f_i}.$$

---

[9] By extension, the analysis applies to formulation (*OPT3*) as well.

6.2 One-Dimensional Types

With this notation,

$$L_1(\lambda) = \max \sum_{i=1}^{m} f_i \mathcal{A}_i h_i(\lambda)$$

$$\text{s.t.} \sum_{i \geq r} f_i \mathcal{A}_i \leq \frac{1 - F(r-1)^n}{n} \quad \forall r \in \{1, 2, \ldots, m\}$$

$$\mathcal{A}_i \geq 0 \;\; \forall i = 1, \ldots, m.$$

We will show that there is an optimal solution λ^* to $\min_{\lambda \geq 0} L_1(\lambda)$ and threshold k^* such that $h_m(\lambda^*) \geq h_{m-1}(\lambda^*) \geq \cdots \geq h_{k^*}(\lambda^*) > 0$ and $h_i(\lambda^*) \leq 0$ for all $i \leq k^* - 1$. In this case, $h_i(\lambda^*)$ is called an **ironed virtual value**.[10] If we replace each $v(i)$ in the objective function of *(OPT6)* by $h_i(\lambda^*)$, we can apply Theorem 6.2.7. This yields the optimal auction when the monotone hazard rate condition does not apply. The optimal mechanism is to award the object to the agent with the highest ironed virtual value provided it is nonnegative. Notice that this produces a monotone expected allocation.

Furthermore, by complementary slackness, $\lambda_i^* > 0$ implies that in an optimal allocation, $\mathcal{A}_i = \mathcal{A}_{i+1}$. Combined with the observation that in each profile the object is awarded with equal probability to the agents with largest nonnegative ironed virtual value, we can describe the form of the optimal allocation. It will consist of a series of cutoffs $\{k_1, k_2, \ldots, k_q\}$ such that

(1) $\mathcal{A}_i = 0$ for all $i \leq k_1$,
(2) $\mathcal{A}_i = \frac{F(k_{j+1}-1)^n - F(k_j-1)^n}{n(F(k_{j+1}-1) - F(k_j-1))}$ for all $i \in [k_j, k_{j+1} - 1]$ and $j = 2, 4, \ldots, 2\lfloor q/2 \rfloor$ and $i \in [k_j, k_{j+1} - 1]$,
(3) $\mathcal{A}_i = \frac{F(i)^n - F(i-1)^n}{nf_i}$ for all $i \in [k_j + 1, k_{j+1} - 1]$ and $j = 1, 3, \ldots, 2\lfloor q/2 \rfloor + 1$.

Call an allocation rule of this form **quasiefficient**. The next result implies that there is a solution to *(OPT6)* that is quasiefficient.

Theorem 6.2.8 *There exists an optimal solution λ^* to $\min_{\lambda \geq 0} L_1(\lambda)$ and threshold k^* such that $h_m(\lambda^*) \geq h_{m-1}(\lambda^*) \geq \cdots \geq h_{k^*}(\lambda^*) > 0$ and $h_i(\lambda^*) \leq 0$ for all $i \leq k^* - 1$.*

Proof. In problem *(OPT6)*, let β_r be the dual variable associated with the constraint (6.10) and λ_i the dual variable associated with the constraint $\mathcal{A}_i \leq$

---

[10] In fact, one can prove that the $h_i(\lambda^*)$ are nondecreasing irrespective of sign. For optimization purposes this is unnecessary.

$\mathcal{A}_{i+1}$. Then, the dual to problem (*OPT6*) is:

$$Z_1 = \min \sum_{r=1}^{m} \frac{1 - F(r-1)^n}{n} \beta_r \qquad (DOPT6)$$

$$\text{s.t.} \sum_{i=1}^{r} \beta_i + \frac{\lambda_r}{f_r} - \frac{\lambda_{r-1}}{f_r} \geq v(r) \ \forall r \in \{1, 2, \ldots, m\}$$

$$\beta_r, \lambda_r \geq 0 \ \forall r$$

Let $\mathcal{A}^*$ be an optimal solution to (*OPT6*) and (β^*, λ^*) and optimal solution to (*DOPT6*). Let $\underline{r}$ be the lowest type for which $\mathcal{A}^*_{\underline{r}} > 0$.

Because $\mathcal{A}^*_r > 0$ for all $r \geq \underline{r}$, complementary slackness implies that

$$\sum_{i=1}^{r} \beta_i^* + \frac{\lambda_r^*}{f_r} - \frac{\lambda_{r-1}^*}{f_r} = v(r) \ \forall r \geq \underline{r}.$$

Rewriting this last system of equations:

$$h_r(\lambda^*) = \sum_{i=1}^{r} \beta_i^* \ \forall r \geq \underline{r}.$$

Because the β_i^* are nonnegative, it follows that $h_r(\lambda^*)$ is nonnegative and nondecreasing for $r \geq \underline{r}$. Thus, there is a threshold $k^* \geq \underline{r}$ that satisfies the requirements of the theorem. It remains to show that $h_r(\lambda^*) \leq 0$ for all $r < \underline{r}$. Now $\mathcal{A}^*_r = 0$ for all $r < \underline{r}$. Therefore,

$$\sum_{i \geq r} f_i \mathcal{A}_i^* < \frac{1 - F(r-1)^n}{n} \ \forall r < \underline{r}.$$

Hence, by complementary slackness, $\beta_r^* = 0$ for all $r < \underline{r}$. Thus,

$$\frac{\lambda_r^*}{f_r} - \frac{\lambda_{r-1}^*}{f_r} \geq v(r) \ \forall r < \underline{r}.$$

In other words, $h_r(\lambda^*) \leq 0$ for all $r < \underline{r}$. ∎

The proof of Theorem 6.2.8 made no particular use of the functional form of the virtual valuation. Hence, in problem (*OPT6*), we could replace each $v(i)$ by any real number and Theorem 6.2.8 would still hold. From the fundamental theorem of linear programming, it would follow that every extreme point of the following system is quasiefficient:

$$\sum_{i \geq r} f_i \mathcal{A}_i \leq \frac{1 - F(r-1)^n}{n} \ \forall r \in \{1, 2, \ldots, m\}. \qquad (6.11)$$

$$0 \leq \mathcal{A}_1 \leq \cdots \leq \mathcal{A}_i \leq \cdots \leq \mathcal{A}_m.$$

This same observation is made in Manelli and Vincent (2008) using a different argument.

6.2 One-Dimensional Types

6.2.5 From Expected Allocations to the Allocation Rule

Here we will show how to derive the allocation rule that corresponds to a quasiefficient expected allocation. As a by-product, we prove that every quasi-efficient expected allocation rule can be implemented in dominant strategies.[11] Because the quasiefficient allocations are extreme points of system (6.11), it follows that every Bayesian incentive compatible auction can be implemented in a dominant strategy way.

Fix a quasiefficient interim allocation $\mathcal{A}'$. Recall that: it will be defined by a sequence of cutoffs $\{k_1, k_2, \ldots, k_q\}$ such that:

(1) $\mathcal{A}'_i = 0$ for all $i \leq k_1$,
(2) $\mathcal{A}'_i = \frac{F(k_{j+1}-1)^n - F(k_j-1)^n}{n(F(k_{j+1}-1) - F(k_j-1))}$ for all $i \in [k_j, k_{j+1} - 1]$ and $j = 2, 4, \ldots, 2\lfloor q/2 \rfloor$,
(3) $\mathcal{A}'_i = \frac{F(i)^n - F(i-1)^n}{n f_i}$ for all $i \in [k_j + 1, k_{j+1} - 1]$ where $j = 1, 3, \ldots, 2\lfloor q/2 \rfloor + 1$.

Assign an agent with type i a priority $\mathcal{A}'_i$. Consider the allocation rule **a** that assigns the good with equal probability among the agents with highest priority. Notice that the actual value of the $\mathcal{A}'_i$'s is not used to determine an allocation, only their relative values. Observe that **a** is monotone in the type of any one agent holding the types of all other agents fixed. Hence, **a** is IDS. It remains to verify that under **a**, the corresponding $\mathcal{A}_i$'s have the correct value. That is, $\mathcal{A}_i = \mathcal{A}'_i$ for all i.

If $i \in [k_j + 1, k_{j+1} - 1]$ where $j = 1, 3, \ldots, 2\lfloor q/2 \rfloor + 1$, then mimicking the argument in the last part of Section 6.2.2 yields $\mathcal{A}_i = \mathcal{A}'_i$.

Suppose, then, $i \in [k_j, k_{j+1} - 1]$ for $j = 2, 4, \ldots, 2\lfloor q/2 \rfloor$. The probability of drawing a profile of types where $b - 1$ agents have the same priority as type i and the other $(n-1) - (b-1)$ agents have lower priorities is

$$\binom{n-1}{b-1} F(k_j - 1)^{n-b} \Big(\sum_{r=k_j}^{k_{j+1}-1} f_r\Big)^{b-1}.$$

Therefore,

$$\mathcal{A}_i = \sum_{b=1}^{n} \frac{1}{b} \binom{n-1}{b-1} F(k_j - 1)^{n-b} \Big(\sum_{r=k_j}^{k_{j+1}-1} f_r\Big)^{b-1}$$

$$= \sum_{b=1}^{n} \frac{1}{b} \binom{n-1}{b-1} F(k_j - 1)^{n-b} (F(k_{j+1} - 1) - F(k_j - 1))^{b-1}$$

$$= \frac{\sum_{b=1}^{n} \binom{n}{b} F(k_j - 1)^{n-b} (F(k_{j+1} - 1) - F(k_j - 1))^{b}}{n(F(k_{j+1} - 1) - F(k_j - 1))}$$

$$= \frac{[F(k_{j+1} - 1)^n - F(k_j - 1)^n]}{n(F(k_{j+1} - 1) - F(k_j - 1))} = \mathcal{A}'_i.$$

---

[11] See Manelli and Vincent (2008). This is not true in general as shown in Gershkov, Moldovanu and Shi (2011).

6.2.6 Correlated Types

In this section, we examine the design of an optimal auction to sell a single object when the types of the bidders are *not* independent.[12] For economy of notation only, we assume two agents.

The case of correlated types is dramatically different from the independent private values case. To illustrate, suppose the type of an agent can be either high (H) or low (L) with equal probability and $H > L$. Further, the types of the two agents are perfectly correlated. Hence, for example, if agent 1's type is H, then agent 2's type is also H. Consider now the following mechanism:

(1) Each agent reports their type.
(2) If the types differ, "shoot" both agents.
(3) If the types coincide, choose one of them with equal probability and hand her the object and charge her her reported type.

Observe that truthfully reporting one's type is an equilibrium. Further, in contrast to the independent type case, the expected surplus of each agent under such an equilibrium is zero. The expected revenue of the auctioneer is $\frac{1}{2}H + \frac{1}{2}L$. Notice that this is the largest possible expected revenue the auctioneer could hope to realize. In the independent type case, the optimal auction generates an expected revenue much less than this. Thus, when types are correlated, the expected revenue of the auctioneer increases. Why is this? Correlation between types reduces the information rents that an agent commands for his or her private information. In our example, when agent 1 knows his type is H, he also knows agent 2's type. The auctioneer can use this fact to extract more surplus from each agent by getting them to report not only their own types but information about the types of other agents as well. These cross-reports can then be checked for consistency.

Recall that t^i is the type or value that agent i places on the object. For any two $u, v \in T$, let $f_{uv} = Pr[t^2 = v | t^1 = u] = Pr[t^1 = v | t^2 = u]$. The important assumption we make is that no row of the matrix $\{f_{uv}\}$ is a nonnegative linear combination of the other rows. We refer to this as the **cone assumption**. Were the types drawn independently, the rows of the matrix $\{f_{uv}\}$ would be identical, violating the cone assumption. What does the cone assumption buy us? It allows us to distinguish between types of agent 1, say, on the basis of their beliefs about agent 2. For example, agent 1 of type u believes that agent 2 has type v with probability f_{uv}. Thus, it is possible to offer agent 1 of type u a lottery based on $\{f_{uv}\}_{v \in T}$, with payments contingent on agent 2's type, that would be rejected by any agent 1 of type $k \neq u$.

Let p^1_{uv} be the payment that agent 1 makes if he reports u and agent 2 reports v. Similarly define p^2_{uv}. Let $\mathbf{a}_1(u, v)$ be the probability that the object is

---

[12] This case was examined with an example in Myerson (1981) and more generally in Cremer and McLean (1988).

6.2 One-Dimensional Types

assigned to agent 1 when agent 1 reports u and agent 2 reports v. Notice that $\mathbf{a}_1(u, v) = 1 - \mathbf{a}_2(u, v)$.

If agent 1's valuation for the object is u, the relevant BNIC constraint will be:

$$\sum_{v \in T} f_{uv}[\mathbf{a}_1(u, v)u - p_{uv}^1] \geq \sum_{v \in T} f_{uv}[\mathbf{a}_1(k, v)u - p_{kv}^1] \quad \forall k \in T \setminus u.$$

The left-hand side of this inequality is the expected payoff (assuming the other agent reports truthfully) to an agent with type u who reports u. The right hand-side is the expected payoff (assuming the other agent reports truthfully) to an agent with type u who reports k. This constraint must hold for each $u \in T$ and a similar one must hold for agent 2.

The individual rationality constraint for agent 1 is:

$$\sum_{v \in T} f_{uv}[\mathbf{a}_1(u, v)u - p_{uv}^1] \geq 0 \ \forall u \in T.$$

A similar constraint holds for agent 2 as well.

Our goal is to design a mechanism that maximizes expected revenue subject to incentive compatibility and individual rationality. Notice that expected revenue is maximized when the expected surplus to all agents is 0. Given incentive compatibility, agent 1's expected surplus when his type is u is

$$\sum_{v \in T} f_{uv}[\mathbf{a}_1(u, v)u - p_{uv}^1].$$

A similar expression holds for agent 2. So, the auctioneer maximizes expected revenue if she can choose $\mathbf{a}_j$ and p^j so that for all $u \in T$ agent 1's expected surplus is zero, that is,

$$\sum_{v \in T} f_{uv}[\mathbf{a}_1(u, v)u - p_{uv}^1] = 0,$$

and agent 2's expected surplus for all $v \in T$ is zero, that is,

$$\sum_{u \in T} f_{vu}[\mathbf{a}_2(u, v)v - p_{uv}^2] = 0.$$

Substituting this into the incentive compatibility and individual rationality constraints yields the following:

$$\sum_{v \in T} f_{uv}[\mathbf{a}_1(k, v)u - p_{kv}^1] \leq 0, \quad \forall k \in T \setminus u, \ u \in T,$$

$$\sum_{u \in T} f_{vu}[\mathbf{a}_2(u, k)v - p_{uk}^2] \leq 0, \quad \forall k \in T \setminus v, \ v \in T,$$

$$\sum_{v \in T} f_{uv}[\mathbf{a}_1(u, v)u - p_{uv}^1] = 0, \quad \forall u \in T,$$

$$\sum_{u \in T} f_{vu}[\mathbf{a}_2(u, v)v - p_{uv}^2] = 0, \quad \forall v \in T.$$

Now, fix the value of $\mathbf{a}_j$ in the inequalities above and determine if there is a feasible p^j. We employ the Farkas Lemma to verify feasibility. This approach does not explicitly reveal the mechanism. However, if we succeed, then *any* allocation rule can be implemented to generate maximum expected revenue. This is the power of the common prior assumption, correlation and the fact that the agents are risk neutral.

Rewriting the above inequalities by moving terms that are fixed to the right-hand side (with a change in index on the last two to make the Farkas alternative easier to write out):

$$-\sum_{v \in T} f_{uv} p^1_{kv} \leq -\sum_{v \in T} f_{uv} \mathbf{a}_1(k,v)u, \quad \forall k \in T \setminus u, \ u \in T,$$

$$-\sum_{u \in T} f_{uv} p^2_{ak} \leq -\sum_{u \in T} f_{uv} \mathbf{a}_2(u,k)v, \quad \forall k \in T \setminus v, \ v \in T,$$

(6.12)

$$\sum_{v \in T} f_{kv} p^1_{kv} = \sum_{v \in T} f_{kv} \mathbf{a}_1(k,v)k, \quad \forall k \in T,$$

$$\sum_{u \in T} f_{ak} p^2_{uk} = \sum_{u \in T} f_{uk} \mathbf{a}_2(u,k)k, \quad \forall k \in T.$$

Let y^1_{uk} be the variable associated with the first inequality, y^2_{kv} be associated with second inequality, z^1_k with the third, and z^2_k with the fourth set of inequalities.

The Farkas Lemma implies there is no solution to (6.12) if there is a solution to the system:

$$-\sum_{u \neq k} f_{uv} y^1_{uk} + f_{kv} z^1_k = 0, \quad \forall k, v \in T,$$

$$-\sum_{v \neq k} f_{uv} y^2_{kv} + f_{uk} z^2_k = 0, \quad \forall u, k \in T,$$

$$y^i_{uv} \geq 0, \quad \forall u, v \in T, i = 1, 2$$

such that

$$-\sum_{u \in T} \sum_{k \neq u} \left[\sum_{v \in T} f_{uv} \mathbf{a}_1(k,v)u \right] y^1_{uk} - \sum_{v \in T} \sum_{k \neq v} \left[\sum_{u \in T} f_{uv} \mathbf{a}_2(u,k)u \right] y^2_{kv}$$

$$+ \sum_{k \in T} \sum_{v \in T} f_{kv} \mathbf{a}_1(k,v)k z^1_k + \sum_{k \in T} \sum_{u \in T} f_{kv} \mathbf{a}_2(k,v)k z^2_k < 0.$$

From the first two equations, nonnegativity of the f's and the y's, we conclude that the z's must be nonnegative as well. The last inequality which is strict, prevents all the y variables being zero. Given this, the first two equations

6.2.7 The Classical Approach

We now compare the approach described in the preceding pages to the standard approach in the case of independent types. The obvious difference is that the classical approach assumes the type of a buyer lies in the interval $[\underline{t}, \bar{t}]$ according to the distribution function $F(t)$. As before, we focus on the case of a seller selling a single object to n buyers whose types are independent draws from the distribution F. A buyer of type t values a probability q of receiving the good as $v(q|t) = tq$.

The central idea of Myerson's (1981) approach is the use of the indirect utility function $U(t) = t\mathcal{A}_t - P_t$. We can then rewrite the BNIC constraints using the indirect utility function as

$$U(t) \geq t\mathcal{A}_{t'} - P_{t'} \quad \forall t, t', \tag{6.13}$$

$$U(t') \geq t'\mathcal{A}_t - P_t \quad \forall t, t'. \tag{6.14}$$

Combining (6.13) and (6.14) gives

$$(t - t')\mathcal{A}_{t'} \leq U(t) - U(t') \leq (t - t')\mathcal{A}_t, \tag{6.15}$$

which implies that $\mathcal{A}_t$ and $U(t)$ are nondecreasing. The monotonicity of $\mathcal{A}_t$ implies that it is *a.e.* continuous. Thus, rewriting (6.15) as

$$\mathcal{A}_{t'} \leq \frac{U(t) - U(t')}{t - t'} \leq \mathcal{A}_t,$$

and taking limits as $t' \to t$ at points where $\mathcal{A}_t$ is continuous, we conclude that $U'(t) = \mathcal{A}_t$ a.e., which in turn immediately gives absolute continuity of $U(t)$, and hence

$$U(t) = U(\underline{t}) + \int_{\underline{t}}^{t} \mathcal{A}_x dx,$$

and

$$P_t = t\mathcal{A}_t - \int_{\underline{t}}^{t} \mathcal{A}_x dx - U(\underline{t}). \tag{6.16}$$

The expression (6.16) is the continuous analog of the discrete expression (6.2). It is as if the payment P_t was set by following a continuous path from the outside option, given by the indirect utility for the lowest type $\underline{t}$, to type t through a continuous set (or graph) $[\underline{t}, \bar{t}]$. It is important to observe that this path represents the shortest path consistent with the binding BNIC constraints along the way (in the continuous case, we interpret this to mean that the BNIC constraints for type t bind in an infinitesimal neighborhood of t). Fortunately, there is only one possible direction for the shortest path between the lowest possible type and type t in one dimension, that is, the path that goes through all intermediate types between the lowest possible type and type t. Thus, there is no conceptual difference between the continuous and discrete approach.

The Border inequalities in the continuous type case are the obvious "smoothing" of the discrete case:

$$n \int_S f(t) A_t dt \leq 1 - (1 - F(S))^n \ \forall S \subseteq [\underline{t}, \overline{t}].$$

Ironing in the model of discrete types is conceptually similar but computationally easier than in the continuous type model. The continuous type case requires the solution of a Hamiltonian problem; in the discrete case it requires a linear program.

From the Discrete to the Continuous

Here we outline how the optimal mechanism for the continuous type space can be obtained as a limit of the discrete type case. Suppose the type space is $T = [0, 1]$ and let $\{\epsilon, 2\epsilon, \ldots, m\epsilon\}$ be a discretization of T where $m\epsilon = 1$. Denote by Q the compact, convex set defined by the Border inequalities in the continuous type case. Let Q_ϵ be the corresponding polyhedron defined by the Border inequalities in the discrete case. Clearly, $P_\epsilon \subset Q$ and $Q_\epsilon \to Q$ as $\epsilon \to 0$. Let x_ϵ denote the optimal mechanism in Q_ϵ and x the optimal mechanism in P. Without loss, we may suppose that x is an extreme point of P and that it is unique. It is easy to show that the optimal objective function value of x_ϵ must converge to the optimal objective function value of x. This only established that the optimal revenues from the discrete and continuous case are close together for small ϵ. The corresponding mechanisms may be very different. However, x_ϵ can be expressed as a convex combination of extreme points of P. For ϵ small, one of these extreme points must have a revenue very close to the revenue of x_ϵ and so, the revenue of x. Thus, as $\epsilon \to 0$, we identify a sequence of extreme points of Q whose revenue converges to the revenue from x. Because x is the unique optimal extreme point, the sequence of extreme points must converge to x and so the same must be true for x_ϵ.

Interdependent Values

Here we outline the revenue-maximizing ex-post incentive compatible auction for a single object when valuations are interdependent. Notation and terminology can be found in Section 4.5. The derivation follows the single-object private values case.

Suppose the presence of a uniformly ordered case and let $\mathbf{a}_k(\mathbf{s})$ be the allocation to agent k when the profile of reported signals is $\mathbf{s}$. The goal is to identify the allocation rule **a** that maximizes

$$\int_{S^n} \sum_{k=1}^{n} V_k(\mathbf{s})\mathbf{a}_k(\mathbf{s}) f(\mathbf{s}) d\mathbf{s}.$$

Here, $V_k(\mathbf{s})$ is the natural analogue of the virtual value. Specifically,

$$V_k(\mathbf{s}) = v^k(\mathbf{s}) - \left(\frac{1 - F(s^k|\mathbf{s}^{-k})}{f(s^k|\mathbf{s}^{-k})}\right)\left(\frac{\partial v^k(\mathbf{s})}{\partial s^k}\right)$$

If we impose the condition that the V_k's satisfy top single crossing, then the optimal auction awards the object at profile **s** to the agent k with largest $V_k(\mathbf{s})$. Now, top single crossing, unlike the monotone hazard condition, is a restriction on the distribution F of signals as well as the valuation functions v^k. Thus, for this result to be useful, one must identify restrictions on the primitives that would lead to the virtual values being top single crossing.[13]

6.3 BUDGET CONSTRAINTS

In this section, we show how the machinery developed earlier in the chapter can be employed to determine the form of the optimal auction of a single object when all bidders have the same common knowledge budget b. That is, no bidder can pay more than b in any scenario.[14] This example is chosen for two reasons. First, the problem was considered by Laffont and Robert (1996) assuming a continuous type space. However, their analysis is not entirely correct. Second, the method used in Section 6.2.2 to go from the optimal $\mathcal{A}$ to the optimal **a** will fail here. This is because the budget constraint will scupper any attempt to decompose the problem profile by profile. The polyhedral approach based on Border's Theorem circumvents these difficulties.

Incorporating the budget constraint means appending to problem $(OPT6)$ the additional constraint

$$P_i = i\mathcal{A}_i - \sum_{j=1}^{i-1} \mathcal{A}_j \leq b \;\forall i.$$

In words, the *expected* payment of any bidder may not exceed b. This may appear weaker that what is required: In all instances, the payment cannot exceed b. It is not. We can think of the mechanism as charging any bidder who reports type i, P_i, irrespective of whether they receive the good or not. Such an approach would violate *ex-post* individual rationality but not the form of the IR constraint assumed here: *Expected* payoff from participation is nonnegative.

---
[13] For an example, see Branco (1996).
[14] This section is based on Pai and Vohra (2008). That paper also considers the case when budgets are private information.

Hence, our problem can be formulated as

$$Z_2 = \max \sum_{i=1}^{m} f_i \mathcal{A}_i v(i) \quad (BOPT)$$

$$\text{s.t.} \sum_{i \geq r} f_i \mathcal{A}_i \leq \frac{1 - F(r-1)^n}{n} \quad \forall r \in \{1, 2, \ldots, m\}$$

$$i \mathcal{A}_i - \sum_{j=1}^{i-1} \mathcal{A}_j \leq b \; \forall i$$

$$0 \leq \mathcal{A}_1 \leq \cdots \leq \mathcal{A}_i \leq \cdots \leq \mathcal{A}_m$$

For convenience, we have neglected to scale the objective function by n, the number of agents. Monotonicity of the $\mathcal{A}_i$'s means that the P_i must be monotone as well. Thus, if a budget constraint in problem ($BOPT$) binds for some type, r, say, it must bind for all types larger than r. Therefore, at optimality, there is an r^* such that $P_j = P_{r^*}$ for all $j \geq r^*$. Further, incentive compatibility requires that $\mathcal{A}_j = \mathcal{A}_{r^*}$ for all $j \geq r^*$. In effect, all types r^* and larger can be lumped together into a single type r^* with virtual value $v(r^*)$ that occurs with probability $1 - F(r^* - 1)$. Hence, if we are given the value of r^*, we can reformulate problem ($BOPT$) as

$$Z_2(r^*) = \max \sum_{i=1}^{r^*-1} f_i \mathcal{A}_i v(i) + [1 - F(r^* - 1)] \mathcal{A}_{r^*} v(r^*)$$

$$\text{s.t.} \sum_{i=r}^{r^*-1} f_i \mathcal{A}_i + [1 - F(r^* - 1)] \mathcal{A}_{r^*}$$

$$\leq \frac{1 - F(r-1)^n}{n} \quad \forall r \in \{1, 2, \ldots, r^* - 1\} \quad (6.17)$$

$$[1 - F(r^* - 1)] \mathcal{A}_{r^*} \leq \frac{1 - F(r^* - 1)^n}{n} \quad (6.18)$$

$$r^* \mathcal{A}_{r^*} - \sum_{j=1}^{r^*-1} \mathcal{A}_j = b \quad (6.19)$$

$$0 \leq \mathcal{A}_1 \leq \cdots \leq \mathcal{A}_i \leq \cdots \leq \mathcal{A}_{r^*}$$

Notice that the budget constraints for types $i < r^*$ have been dropped. This is because montonicity of payments means they are implied by the budget constraint for type r^*. Also, the formulation assumes that the budget constraint binds at optimality. If not, we could drop it altogether and return to the case with no budgets.

6.3 Budget Constraints

Now take the dual of the program that defines $Z_2(r^*)$. Denote the dual variable associated with (6.17) and (6.18) by β_r and the dual variable associated with (6.19) by η. Let λ_i be the dual variable associated with the constraint $\mathcal{A}_i - \mathcal{A}_{i+1} \leq 0$. By the duality theorem,

$$Z_2(r^*) = \min_{\eta, \{\beta_r\}_1^{r^*}, \{\lambda_r\}_1^{r^*-1}} b\eta + n^{-1} \sum_{r=1}^{r^*} [1 - F(r-1)^n]\beta_r$$

$$r^*\eta + [1 - F(r^* - 1)] \sum_{i=1}^{r^*} \beta_i - \lambda_{r^*-1} \geq [1 - F(r^* - 1)]v(r^*))$$

$$-\eta + f_r \sum_{i=1}^{r} \beta_i + \lambda_r - \lambda_{r-1} \geq f_r v(r) \quad \forall r \leq r^* - 1$$

$$\beta_r, \lambda_r \geq 0 \; r = \{1, \ldots, r^*\}.$$

Let $\mathcal{A}^*$ be an optimal solution to the program that defines $Z_2(r^*)$. Let $\underline{r}$ be the lowest type for which $\mathcal{A}^*_{\underline{r}} > 0$. Complementary slackness implies that:

$$r^*\eta + (1 - F(r^* - 1)) \sum_{i=1}^{r^*} \beta_i - \lambda_{r^*-1} = [1 - F(r^* - 1)]v(r^*) \tag{6.20}$$

$$-\eta + f_r \sum_{i=1}^{r} \beta_i + \lambda_r - \lambda_{r-1} = f_r v(r) \quad \forall \underline{r} \leq r \leq r^-1*. \tag{6.21}$$

Rewriting (6.20–6.21) yields the following upper triangular system:

$$\sum_{i=1}^{r^*} \beta_i - \frac{\lambda_{r^*-1}}{1 - F(r^* - 1)} = v(r^*) - r^* \frac{\eta}{1 - F(r^* - 1)},$$

$$\sum_{i=1}^{r} \beta_i + \frac{\lambda_r}{f_r} - \frac{\lambda_{r-1}}{f_r} = v(r) + \frac{\eta}{f_r} \; \forall \underline{r} \leq r \leq r^* - 1.$$

Intuitively, these equations tell us that the "correct" virtual valuation of a type r should be $v(r) + \frac{\eta}{f_r}$. Here, $\frac{\eta}{f_r}$ corrects for the budget constraint: Allocating to lower types reduces the payment of the high types, and hence "relaxes" the budget constraint.

By analogy with the monotone hazard rate condition, we could assume that the corrected virtual values, $v(r) + \frac{\eta}{f_r}$, are monotone. A sufficient condition for this is that f_r is decreasing and satisfies the monotone hazard rate condition.[15] By analogy with the unbudgeted case, there will be a lowest type, $\underline{r}$, that will be allocated $\underline{r}$ will be the smallest value of r for which $v(r) + \frac{\eta}{f_r} \geq 0$. Finally, the optimal allocation rule will coincide with the efficient allocation rule between types $r^* - 1$ and $\underline{r}$.

---

[15] This is a severe restriction.

Theorem 6.3.1 *Suppose f_i is decreasing in i and F satisfies the monotone hazard rate condition, that is, $v(i)$ is increasing in i. Then, the optimal solution $\mathcal{A}^*$ of (BOPT) can be described in terms of two cutoffs $r^* > \underline{r}$:*

(1) $\mathcal{A}_i^* = 0$ for all $i \leq \underline{r}$.
(2) $\mathcal{A}_i^* = \mathcal{A}_{r^*}^*$ for all $i \geq r^*$.
(3) The allocation rule will be efficient between types $r^* - 1$ and $\underline{r}$.
(4) $\underline{r}$ is the lowest type such that

$$v(\underline{r}) + \frac{\eta}{f_{\underline{r}}} \geq 0,$$

where

$$\eta = \frac{(1 - F(r^* - 1))(1 - F(r^* - 2))}{r^* f_{r^*-1} + (1 - F(r^* - 1))}.$$

Proof. We proceed by constructing dual variables that complement the primal solution described in the statement of the Theorem.

Since $\mathcal{A}_i^* = 0$ for $i < \underline{r}$ and $f_i > 0$ for all i, the Border constraints for $i < \underline{r}$ do not bind at optimality. Therefore, $\beta_i = 0$ for all $i < \underline{r}$. Further, $0 = \mathcal{A}_{\underline{r}-1}^* < \mathcal{A}_{\underline{r}}^*$ by definition of $\underline{r}$, and so, to ensure complementary slackness, we need $\lambda_{\underline{r}-1} = 0$. Similarly, because r^* is the lowest type for which the budget constraint binds, $\mathcal{A}_{r^*-1}^* < \mathcal{A}_{r^*}^*$, requiring that $\lambda_{r^*-1} = 0$.

Subtracting the dual constraints corresponding to types r^* and $r^* - 1$ and using the fact that $\lambda_{r^*-1} = 0$, we have:[16]

$$\beta_{r^*} + \frac{\lambda_{r^*-2}}{f_{r^*-1}} = v(r^*) - r^* \frac{\eta}{1 - F(r^* - 1)} - v(r^* - 1) - \frac{\eta}{f_{r^*-1}}. \quad (6.22)$$

Subtracting the dual constraints corresponding to i and $i - 1$, where $\underline{r} + 1 \leq i \leq r^* - 1$, we have:

$$\beta_i + \frac{\lambda_i}{f_i} - \frac{\lambda_{i-1}}{f_i} - \frac{\lambda_{i-1}}{f_{i-1}} + \frac{\lambda_{i-2}}{f_{i-1}} = v(i) + \frac{\eta}{f_i} - v(i - 1) - \frac{\eta}{f_{i-1}}. \quad (6.23)$$

Finally, the dual constraint corresponding to type $\underline{r}$ reduces to:

$$\beta_{\underline{r}} + \frac{\lambda_{\underline{r}}}{f_{\underline{r}}} = v(\underline{r}) + \frac{\eta}{f_{\underline{r}}} \quad (6.24)$$

It suffices to identify a nonnegative solution to the system (6.22–6.24) such that $\beta_i = 0$ for all $i < \underline{r}$ and $\lambda_{\underline{r}-1} = 0$.

Consider the following solution:

$$\beta_{r^*} = 0 \quad (6.25)$$

$$\beta_i = v(i) - v(i - 1) + \eta\left(\frac{1}{f_i} - \frac{1}{f_{i-1}}\right) \quad \underline{r} + 1 \leq i \leq r^* - 1 \quad (6.26)$$

$$\lambda_{i-1} = 0 \quad \underline{r} + 1 \leq i \leq r^* - 1 \quad (6.27)$$

$$\eta = \frac{(1 - F(r^* - 1))(1 - F(r^* - 2))}{r^* f_{r^*-1} + (1 - F(r^* - 1))} \quad (6.28)$$

---

[16] This step is where the upper triangular constraint matrix is helpful.

6.3 Budget Constraints

Direct computation verifies that the given solution satisfies (6.22–6.24). In fact, it is the unique solution to (6.22–6.24) with all λ's equal to zero. All variables are nonnegative. In particular, β_v for $\underline{r}+1 \leq v \leq r^*-1$ is positive. This is because F satisfies the monotone hazard rate and decreasing density conditions; hence, for any i, $v(i) - v(i-1) + \eta(\frac{1}{f_i} - \frac{1}{f_{i-1}}) > 0$. Furthermore, it complements the primal solution described in the statement of Theorem 6.3.1. This concludes the case where our regularity condition on the distribution of types (monotone hazard rate, decreasing density) are met. ∎

Suppose monotonicity of the corrected virtual values fails. That is, $v(i) - v(i-1) + \eta(\frac{1}{f_i} - \frac{1}{f_{i-1}}) < 0$ for some i. The dual solution identified above will be infeasible because $\beta_i < 0$. More generally, there can be no dual solution that satisfies (6.22–6.24) with all $\lambda_i = 0$. Hence, there must be at least one i between $\underline{r}$ and r^*-1 such that $\lambda_i > 0$. This implies, by complementary slackness, that the corresponding primal constraint $\mathcal{A}_i - \mathcal{A}_{i+1} \leq 0$ binds at optimality. In particular to solve the problem, one must carry out an ironing approach of the type described in Theorem 6.2.8.

6.3.1 The Continuous Type Case

Here we examine the same problem but in a continuous type space to provide contrast. Suppose types are drawn from [0, 1] according to the density $f(t)$ and $F(t) = \int_0^t f(t)dt$. Their virtual valuation is defined as $v(t) = t - \frac{1-F(t)}{f(t)}$. The expected allocation to a type t will be $\mathcal{A}_t$.

Assuming the budget constraint binds, we can argue, as in the discrete type space, that we have two cutoffs, t_1, t_2. The allocation rule does not allot to types below t_1, and pools all types above t_2. Because the budget constraint binds for all types t_2 and above:

$$t_2 \mathcal{A}_{t_2} - \int_{t_1}^{t_2} \mathcal{A}_t dt = b. \tag{6.29}$$

Therefore, our problem reduces to finding $\mathcal{A}_t$, t_1 and t_2 to solve:

$$\max_{t_1, t_2, \mathcal{A}_t} \mathcal{A}_{t_2}(1 - F(t_2))t_2 + \int_{t_1}^{t_2} v(t) f(t) \mathcal{A}_t dt \quad (LR)$$

$$\text{s.t. } t_2 \mathcal{A}_{t_2} - \int_{t_1}^{t_2} \mathcal{A}_t dt = b$$

$$n \int_s^1 f(t) \mathcal{A}_t dt \leq 1 - F(s)^n \quad \forall t \in [t_1, 1]$$

$$\mathcal{A}_t = \mathcal{A}_{t_2} \quad \forall t \geq t_2$$

$$\mathcal{A}_t \text{ monotone.}$$

One might conjecture, as in the discrete type case, that in the interval $[t_1, t_2]$, the optimal allocation assigns the good to the highest type. This will make the allocation rule:

$$\mathcal{A}_t = \begin{cases} \frac{1-F^n(t_2)}{n(1-F(t_2))} & t \geq t_2 \\ F^{n-1}(t) & t \in [t_1, t_2] \\ 0 & o.w. \end{cases} \tag{6.30}$$

Verifying that (6.30) is optimal would require taking the dual of the infinite dimensional linear program, that is, defining an appropriate measure on the dual space.[17]

6.4 ASYMMETRIC TYPES

Here we outline very briefly how to accommodate the case when types are drawn from the same set T independently but from different distributions. Suppose agent k's type is drawn from the distribution F^k (with density f^k). Denote by $\mathbf{a_k}(t^k, \mathbf{t}^{-k})$ the allocation to agent i when she reports type t^k and the other agents report $\mathbf{t}^{-k}$. Let $\mathcal{A}_{t^k}^k$ be the expected allocation to agent k when she reports t^k and $P_{t^k}^k$ her expected payment.

Incentive compatibility will still yield monotonicity of the expected allocations. A payment formula of the form in equation (6.2) will still obtain. The objective function will be linear in the expected allocations with coefficients corresponding to virtual values. The only thing that does change are the relevant Border inequalities. With asymmetric types, the Border inequalities (assuming monotonicity) will be (see Border 2007)

$$\sum_{k=1}^{n}\sum_{t^k \geq r^k} f_{t^k}^k \mathcal{A}_{t^k}^k \leq 1 - \Pi_{k=1}^n F^k(r^k) \; \forall r^k \in T.$$

The Border inequalities (after a suitable substitution) still describe a polymatroid, so the previous techniques apply.

6.4.1 Bargaining

An immediate application is to the problem of bargaining between agents in the shadow of private information, a problem considered by Myerson and Satterthwaite (1983). We have a seller with a good and her type is the opportunity cost of the good. Thus we can suppose the seller's type is a draw from $\{-m, -(m-1), \ldots, -2, -1\}$ according to distribution F^1. There is also a buyer whose type is his marginal value for the good. This is a draw from $\{1, 2, \ldots, m\}$ with distribution F^2.

Consider a broker who intermediates the trade. The broker is interested in an individually rational BINC mechanism that maximizes his gain from exchange. The broker will announce an allocation rule and a payment rule. We interpret $\mathbf{a}_1(t^1, t^2)$ as either the probability that the good is taken from the seller or the fraction of the good yielded by the seller when seller reports t^1 and buyer reports t^2. Likewise, $\mathbf{a}_2(t^1, t^2)$ is the probability that the buyer receives the good. Feasibility requires that

$$\mathbf{a}_1(t^1, t^2) + \mathbf{a}_2(t^1, t^2) \leq 1 \; \forall t^1, t^2.$$

---

[17] See, for example, Anderson and Nash (1987).

Expected allocations are defined in the obvious way. Expected payments for each agent as a function of the expected allocation can be derived in the usual way. Thus, as before, we can eliminate the payment variables and, using the Border inequalities for asymmetric types, derive a formulation similar to (OPT6).

6.5 MULTIDIMENSIONAL TYPES

We turn our attention to the case when types are multidimensional. First, we examine where the classical approach, which assumes a continuous type space, would lead us. For completeness, the discussion will repeat some of what is to be found in Section 4.4.

Suppose agents have multidimensional types $i = \{i_1, \ldots, i_D\}$, which are distributed on T, a convex compact subset of $\mathbb{R}_+^D$, with a continuous density function $f(t)$ with $\text{supp}(f) = T$. Valuations of agents are given by $v(\mathbf{a}|i) = i \cdot \mathbf{a} = \sum_{j=1}^{D} i_j \mathbf{a}_j$, where allocations $\mathbf{a} = \{\mathbf{a}_1, \ldots, \mathbf{a}_D\}$ belong to $\mathbb{R}^D$. The indirect utility function is $U(i) = i \cdot \mathcal{A}_i - P_i$. As in the one-dimensional case, we can rewrite the BNIC constraints using the indirect utility function as

$$U(i) \geq i \cdot \mathcal{A}_{i'} - P_{i'} \quad \forall i, i', \tag{6.31}$$

$$U(i') \geq i' \cdot \mathcal{A}_i - P_i \quad \forall i, i'. \tag{6.32}$$

Combining (6.31) and (6.32) gives

$$(i - i') \cdot \mathcal{A}_{i'} \leq U(i) - U(i') \leq (i - i') \cdot \mathcal{A}_i. \tag{6.33}$$

In an abuse of notation, let us rewrite (6.33) as

$$\mathcal{A}_{i'} \leq \frac{U(i) - U(i')}{i - i'} \leq \mathcal{A}_i. \tag{6.34}$$

Suppose $U(i)$ were differentiable almost everywhere. Then, taking limits in (6.34) as $\Delta(i' - i)_d \to 0$ in each coordinate d, and assuming $\mathcal{A}_i$ to be continuous yields $\nabla U(i) = \mathcal{A}_i$,

$$U(i) = U(\underline{i}) + \int_l \mathcal{A}_v dv, \tag{6.35}$$

and

$$P_i = i \mathcal{A}_i - \int_l \mathcal{A}_v dv - U(\underline{i}). \tag{6.36}$$

for any path l in T connecting any two types $\underline{i}$ and i. Hence, the choice of the path to type i through T needed to determine the payment P_i does not matter. However, differentiability of $U(i)$ in the multidimensional case cannot be guaranteed.[18] Hence, the expression (6.36) can not be viewed as a multidimensional analog of (6.16).

---

[18] See Example 1 in Krishna and Perry (2000) for a nondifferentiable indirect utility function.

Notice that differentiability of $U(i)$ is not required to obtain partial derivatives *a.e.*, and so form the gradient. Indeed, taking limits in (6.34) as $(i' - i)_d \to 0$ in each dimension d yields $\nabla U(i) = \mathcal{A}_i$. Unfortunately, the existence of partial derivatives does not guarantee differentiability of $U(i)$.

Rochet and Choné (1998) tell us there exists an expected allocation and price schedules such that the BNIC constrains are satisfied for almost all i if and only if:

(1) $\mathcal{A}_i = \nabla U(i)$ for a.e. i in T,
(2) U is convex and continuous on T.

Notice that $\mathcal{A}_i = \nabla U(i)$ does not imply differentiability of $U(i)$, hence the only analog for expression (6.36) is

$$P_i = i \cdot \nabla U(i) - U(i). \tag{6.37}$$

The discrete multidimensional case has similar difficulties. The payment P_i for a particular allocation scheme is given by the shortest path toward the vertex i (there is no analog to the gradient of the indirect utility function in the multidimensional discrete case).[19] Indeed, the incremental indirect utility $\Delta U(i)$ at the vertex i is determined by the edge (i', i) originating in one of adjacent vertices i', along which the BNIC constraint binds. This binding BNIC constraint also gives the direction of the increase in the indirect utility, but this direction cannot be averaged out to form an average direction of the indirect utility increase (i.e., the indirect utility gradient) due to the discrete nature of this direction.

This logic shows that the indirect utility approach does not work in the discrete multidimensional case, and we have to rely on an enumeration of all the binding BNIC constraints and the corresponding shortest paths through the multidimensional lattice in order to find a solution to the optimal auction problem.

Identifying an optimal mechanism in a multidimensional environment is a hard problem in either the discrete or continuous case.[20] Not surprisingly, many results in which there is an explicit solution to a multidimensional optimal mechanism rely on some properties of the problem that reduce it to a one-dimensional case (Wilson 1993), or reduce the set of available options as in Malakhov and Vohra (2009).[21] We examine an instance of each.[22]

---

[19] Notice that the BNIC constraints bind along the shortest path, which can be interpreted as having the shortest path at each vertex follow the direction of the BNIC constraint that binds at that vertex.
[20] See, for example, Manelli and Vincent (2007). They show that in the multigood setting, the optimal mechanism will typically involve randomization.
[21] Another examples of the second variety is Beaudry, Blackorby, and Szalay (2009).
[22] This section is based on joint work with Alexey Malakhov.

6.5 Multidimensional Types

6.5.1 Wilson's Example

One case where it is possible to find the optimal paths is in a discrete two-dimensional analog of a continuous model first solved by Wilson (1993, Chapter 13). In Wilson's model, types are uniformly distributed on $T = \{t \in R_+^2 : t_1^2 + t_2^2 \leq 1\}$. The utility of an agent with type t for a vector q of goods is $v(q|t) = q \cdot t$. The seller incurs a cost $C(q) = \frac{\|q\|^2}{2}$ to produce a vector q of goods. The objective is to identify the profit-maximizing BNIC mechanism:

$$\max_{\{q,p\}} \int_T P(t) - C(q(t))dt \qquad \text{(OPT-W)}$$

s.t. $v(q(t)|t) - P(t) \geq v(q(s)|t) - P(s) \quad \forall t, s \in T$ (BNIC)

The solution to Wilson's problem is given by

$$q^*(t) = \frac{1}{2} \max\left(0, 3 - \frac{1}{\|t\|^2}\right) t. \qquad (6.38)$$

Discrete Approach to the Problem

We solve problem $(OPT - W)$ in discrete polar coordinates using the network representation. Consider a discrete grid in polar coordinates (r, φ), that is,

$t_1 = r \cos \varphi,$

$t_2 = r \sin \varphi,$

with $r \in \{r_1, \ldots, r_n\}$, where $r_i = \frac{i}{n}$, $r_0 = 0$ and $\varphi \in \{0, \frac{\pi}{2k}, \frac{2\pi}{2k}, \ldots, \frac{\pi}{2}\}$. Consider the direct mechanism approach with the allocation schedule given by

$q_1(r, \varphi) = R(r, \varphi) \cos \theta(r, \varphi),$

$q_2(r, \varphi) = R(r, \varphi) \sin \theta(r, \varphi).$

The BNIC constraints for all (r, φ) and (r', φ') are

$R(r, \varphi) \cos \theta(r, \varphi) r \cos \varphi + R(r, \varphi) \sin \theta(r, \varphi) r \sin \varphi - P(r, \varphi)$

$\geq R(r', \varphi') \cos \theta(r', \varphi') r \cos \varphi + R(r', \varphi') \sin \theta(r', \varphi') r \sin \varphi - P(r', \varphi').$

The cost function is given by

$$C(q) = \frac{R(r, \varphi)^2}{2}.$$

Our approach will be to conjecture that the optimal paths must be radial and then compute an optimal allocation for such a conjecture. This amounts to relaxing some of the BNIC constraints. We complete the argument by showing that the solution found satisfies the relaxed BNIC constraints.

Lemma 6.5.1 *If a payment $P(r_i, \varphi)$ is determined by a radial path*

$(0, \varphi) \to (r_1, \varphi) \to \cdots \to (r_i, \varphi),$

then the optimal allocations are given by

$$q_1(r_i, \varphi) = R(r_i, \varphi)\cos\varphi,$$
$$q_2(r_i, \varphi) = R(r_i, \varphi)\sin\varphi,$$

and the profit-maximizing payment $P(r_i, \varphi)$ is given by

$$P(r_i, \varphi) = \sum_{j=1}^{i} r_j \left[R(r_j, \varphi) - R(r_{j-1}, \varphi) \right]. \tag{6.39}$$

Proof. Suppressing the dependence on φ, which we can, the argument mimics the proofs of Theorem 6.2.2 and Theorem 6.2.3. For completeness, we include a proof below. For variety, we use an argument by induction.

The proof is by induction on $r_i \in \{r_1, \ldots, r_n\}$. First, consider the case of $i = 1$ and $\varphi \in \{0, \frac{\pi}{2k}, \frac{2\pi}{2k}, \ldots, \frac{\pi}{2}\}$.

If the payment is set through a radial path, that is, $(0, \varphi) \longrightarrow (r_1, \varphi)$, then

$$P(r_1, \varphi) = R(r_1, \varphi)\cos\theta(r_1, \varphi_j)(r_1\cos\varphi)$$
$$+ R(r_1, \varphi)\sin\theta(r_1, \varphi)(r_1\sin\varphi),$$

and the profit is

$$\Pi_1 = R(r_1, \varphi)\cos\theta(r_1, \varphi)(r_1\cos\varphi)$$
$$+ R(r_1, \varphi)\sin\theta(r_1, \varphi)(r_1\sin\varphi) - \frac{1}{2}R(r_1, \varphi)^2.$$

Notice that Π_1 is maximized when $\theta(r, \varphi) = \varphi$ (indeed, it does not affect the cost, while maximizing the revenue), hence

$$q_1(r_1, \varphi) = R(r_1, \varphi)\cos\varphi,$$
$$q_2(r_1, \varphi) = R(r_1, \varphi)\sin\varphi,$$
$$\Pi_1 = r_1 R(r_1, \varphi) - \frac{1}{2} R(r_1, \varphi)^2,$$

and

$$P(r_1, \varphi) = r_1 R(r_1, \varphi).$$

Now examine the transition from i to $i + 1$. Assuming that

$$q_1(r_i, \varphi) = R(r_i, \varphi)\cos\varphi,$$
$$q_2(r_i, \varphi) = R(r_i, \varphi)\sin\varphi,$$
$$P(r_i, \varphi) = \sum_{j=1}^{i} r_j [R(r_j, \varphi) - R(r_{j-1}, \varphi)],$$

6.5 Multidimensional Types

it follows that if $P(r_{i+1}, \varphi)$ is determined by the path $(0, \varphi_j) \to \cdots \to (r_i, \varphi) \to (r_{i+1}, \varphi)$, we conclude

$$P(r_{i+1}, \varphi) = R(r_{i+1}, \varphi) \cos \theta(r_{i+1}, \varphi)(r_{i+1} \cos \varphi)$$
$$+ R(r_{i+1}, \varphi) \sin \theta(r_{i+1}, \varphi)(r_{i+1} \sin \varphi)$$
$$- r_{i+1} R(r_i, \varphi) + P(r_i, \varphi),$$

and the profit along the path is

$$\Pi_{i+1} = R(r_{i+1}, \varphi) \cos \theta(r_{i+1}, \varphi)(r_{i+1} \cos \varphi)$$
$$+ R(r_{i+1}, \varphi) \sin \theta(r_{i+1}, \varphi)(r_{i+1} \sin \varphi)$$
$$- r_{i+1} R(r_i, \varphi) + P(r_i, \varphi) - \frac{1}{2} R(r_{i+1}, \varphi)^2$$
$$+ \sum_{l=0}^{i} \left[P(r_l, \varphi) - \frac{1}{2} R(r_l, \varphi)^2 \right]$$

By the same argument as in the case of r_1, we conclude that the above profit is maximized when $\theta(r_{i+1}, \varphi) = \varphi$, hence

$$q_1(r_{i+1}, \varphi) = R(r_{i+1}, \varphi) \cos \varphi,$$
$$q_2(r_{i+1}, \varphi) = R(r_{i+1}, \varphi) \sin \varphi,$$

and

$$P(r_{i+1}, \varphi) = \sum_{j=1}^{i+1} r_j [R(r_j, \varphi) - R(r_{j-1}, \varphi)]. \quad \blacksquare$$

The following is now immediate from Lemma 6.5.1.

Lemma 6.5.2 *If all profit-maximizing payments $P(r_i, \varphi)$ are determined by radial paths*

$$(0, \varphi) \to (r_1, \varphi) \to \cdots \to (r_i, \varphi),$$

then, in an optimal allocation, $R(r_i, \varphi)$ is independent of φ, that is, $R(r_i, \varphi) = R(r_i)$. Therefore,

$$q_1(r_i, \varphi) = R(r_i) \cos \varphi, \tag{6.40}$$
$$q_2(r_i, \varphi) = R(r_i) \sin \varphi. \tag{6.41}$$

Furthermore, optimal payments $P(r_i, \varphi)$ do not depend on φ, that is,

$$P(r_i, \varphi) = P(r_i) = \sum_{l=1}^{i} r_l [R(r_l) - R(r_{l-1})]. \tag{6.42}$$

Lemma 6.5.3 *All profit-maximizing payments $P(r_i, \varphi)$ are determined by radial paths*

$$(0, \varphi) \to (r_1, \varphi) \to \cdots \to (r_i, \varphi).$$

Proof. The proof is by induction on r_i. First, consider the case $i = 1$, and $\varphi \in \{0, \frac{\pi}{2k}, \frac{2\pi}{2k}, \ldots, \frac{\pi}{2}\}$. Denote payments that are determined by radial paths $(0, \varphi) \longrightarrow (r_1, \varphi)$ as $P_r(r_1, \varphi)$, and payments that are determined by nonradial paths $(0, \varphi') \longrightarrow (r_1, \varphi') \longrightarrow (r_1, \varphi)$ as $P_{nr}(r_1, \varphi)$. Then

$$P_r(r_1, \varphi) = R(r_1, \varphi) \cos \theta(r_1, \varphi)(r_1 \cos \varphi)$$
$$+ R(r_1, \varphi) \sin \theta(r_1, \varphi)(r_1 \sin \varphi), \qquad (6.43)$$

and

$$P_{nr}(r_1, \varphi) = R(r_1, \varphi) \cos \theta(r_1, \varphi)(r_1 \cos \varphi)$$
$$+ R(r_1, \varphi) \sin \theta(r_1, \varphi)(r_1 \sin \varphi)$$
$$- R(r_1, \varphi') \cos \theta(r_1, \varphi')(r_1 \cos \varphi)$$
$$- R(r_1, \varphi') \sin \theta(r_1, \varphi')(r_1 \sin \varphi) + P_r(r_1, \varphi').$$

Lemma 6.5.1 implies that $P_r(r_1, \varphi') = r_1 R(r_1, \varphi')$, hence,

$$P_{nr}(r_1, \varphi) = R(r_1, \varphi) \cos \theta(r_1, \varphi)(r_1 \cos \varphi)$$
$$+ R(r_1, \varphi) \sin \theta(r_1, \varphi)(r_1 \sin \varphi)$$
$$- R(r_1, \varphi') \cos \theta(r_1, \varphi')(r_1 \cos \varphi)$$
$$- R(r_1, \varphi') \sin \theta(r_1, \varphi')(r_1 \sin \varphi) + r_1 R(r_1, \varphi'). \quad (6.44)$$

Combining (6.43) and (6.44), we obtain

$$P_{nr}(r_1, \varphi) = P_r(r_1, \varphi) + r_1 R(r_1, \varphi')(1 - \cos \theta(r_1, \varphi') \cos \varphi$$
$$- \sin \theta(r_1, \varphi') \sin \varphi).$$

Finally, since $\cos \theta \cos \varphi + \sin \theta \sin \varphi < 1$ for $\forall \theta \neq \varphi$, we conclude that

$$P_{nr}(r_1, \varphi) \geq P_r(r_1, \varphi),$$

which proves that the radial path is the shorter one, and because payments are determined by shortest paths, profit-maximizing payments $P(r_1, \varphi)$ are determined by radial paths.

Now examine the transition from i to $i + 1$. Assuming that $P(r_i, \varphi)$ are determined by radial paths, Lemma 6.5.2 implies that

$$q_1(r_i, \varphi_j) = R(r_i) \cos \varphi, \qquad (6.45)$$

$$q_2(r_i, \varphi) = R(r_i) \sin \varphi, \qquad (6.46)$$

$$P(r_i, \varphi) = P(r_i) = \sum_{l=1}^{i} r_l \left[r_l R(r_l) - R(r_{l-1}) \right]. \qquad (6.47)$$

6.5 Multidimensional Types

We now show that the payment $P_r(r_{i+1}, \varphi)$, determined by the radial path $(r_i, \varphi) \longrightarrow (r_{i+1}, \varphi)$, is smaller than payments $P_{nr1}(r_{i+1}, \varphi)$, $\{P_{nrk}(r_{i+1}, \varphi)\}_{k \leq i}$, that are determined by paths $(r_i, \varphi') \to (r_{i+1}, \varphi') \to (r_{i+1}, \varphi)$ and $(r_k, \varphi') \to (r_{i+1}, \varphi)$, respectively. Then, using (6.45), (6.46), and (6.47), and without assuming anything about allocations at (r_{i+1}, φ), we get

$$P_r(r_{i+1}, \varphi) = R(r_{i+1}, \varphi) \cos \theta(r_{i+1}, \varphi)(r_{i+1} \cos \varphi)$$
$$+ R(r_{i+1}, \varphi) \sin \theta(r_{i+1}, \varphi)(r_{i+1} \sin \varphi)$$
$$- r_{i+1} R(r_i) + P_r(r_i). \tag{6.48}$$

For the $P_{nr1}(r_{i+1}, \varphi)$, we have

$$P_{nr1}(r_{i+1}, \varphi) = R(r_{i+1}, \varphi) \cos \theta(r_{i+1}, \varphi)(r_{i+1} \cos \varphi)$$
$$+ R(r_{i+1}, \varphi) \sin \theta(r_{i+1}, \varphi)(r_{i+1} \sin \varphi)$$
$$- R(r_{i+1}, \varphi') \cos \theta(r_{i+1}, \varphi')(r_{i+1} \cos \varphi)$$
$$- R(r_{i+1}, \varphi') \sin \theta(r_{i+1}, \varphi')(r_{i+1} \sin \varphi)$$
$$+ P_r(r_{i+1}, \varphi'). \tag{6.49}$$

Now notice that for the $P_{nr1}(r_{i+1}, \varphi)$ to be determined by the shortest path, $P_r(r_{i+1}, \varphi')$ must be determined by the radial path, hence by Lemma 6.5.1,

$$P_r(r_{i+1}, \varphi') = r_{i+1} R(r_{i+1}, \varphi') - r_{i+1} R(r_i) + P_r(r_i). \tag{6.50}$$

Combining (6.49) and (6.50), we obtain

$$P_{nr1}(r_{i+1}, \varphi) = R(r_{i+1}, \varphi) \cos \theta(r_{i+1}, \varphi)(r_{i+1} \cos \varphi)$$
$$+ R(r_{i+1}, \varphi) \sin \theta(r_{i+1}, \varphi)(r_{i+1} \sin \varphi)$$
$$- R(r_{i+1}, \varphi') \cos \theta(r_{i+1}, \varphi')(r_{i+1} \cos \varphi)$$
$$- R(r_{i+1}, \varphi') \sin \theta(r_{i+1}, \varphi')(r_{i+1} \sin \varphi)$$
$$+ r_{i+1} R(r_{i+1}, \varphi') - r_{i+1} R(r_i) + P_r(r_i). \tag{6.51}$$

Finally, (6.48) and (6.51) give

$$P_{nr1}(r_{i+1}, \varphi) = P_r(r_{i+1}, \varphi) + r_{i+1} R(r_{i+1}, \varphi')$$
$$(1 - \cos \theta(r_{i+1}, \varphi') \cos \varphi - \sin \theta(r_{i+1}, \varphi') \sin \varphi),$$

and because $\cos \theta \cos \varphi + \sin \theta \sin \varphi < 1$ for $\forall \theta \neq \varphi$, we conclude that

$$P_{nr1}(r_{i+1}, \varphi) \geq P_r(r_{i+1}, \varphi). \tag{6.52}$$

For $P_{nri}(r_{i+1}, \varphi)$, we have

$$P_{nri}(r_{i+1}, \varphi) = R(r_{i+1}, \varphi) \cos \theta(r_{i+1}, \varphi)(r_{i+1} \cos \varphi)$$
$$+ R(r_{i+1}, \varphi) \sin \theta(r_{i+1}, \varphi)(r_{i+1} \sin \varphi)$$
$$- R(r_i, \varphi') \cos \theta(r_i, \varphi')(r_{i+1} \cos \varphi)$$
$$- R(r_i, \varphi') \sin \theta(r_i, \varphi')(r_{i+1} \sin \varphi) + P_r(r_i, \varphi'). \tag{6.53}$$

Then using (6.45), (6.46), and (6.47), and without assuming anything about allocations at (r_{i+1}, φ), we get

$$P_{nri}(r_{i+1}, \varphi) \geq R(r_{i+1}, \varphi) \cos \theta(r_{i+1}, \varphi)(r_{i+1} \cos \varphi)$$
$$+ R(r_{i+1}, \varphi) \sin \theta(r_{i+1}, \varphi)(r_{i+1} \sin \varphi)$$
$$- r_{i+1} R(r_i) + P_r(r_i). \quad (6.54)$$

Finally, (6.48) and (6.54) give

$$P_{nri}(r_{i+1}, \varphi) \geq P_r(r_{i+1}, \varphi). \quad (6.55)$$

The case of $P_{nrk}(r_{i+1}, \varphi)$ for $k < i$ is subsumed by the previous case because

$$P_{nrk}(r_{i+1}, \varphi) = R(r_{i+1}, \varphi) \cos \theta(r_{i+1}, \varphi)(r_{i+1} \cos \varphi)$$
$$+ R(r_{i+1}, \varphi) \sin \theta(r_{i+1}, \varphi)(r_{i+1} \sin \varphi)$$
$$- R(r_k, \varphi') \cos \theta(r_k, \varphi')(r_{i+1} \cos \varphi)$$
$$- R(r_k, \varphi') \sin \theta(r_k, \varphi')(r_{i+1} \sin \varphi)$$
$$+ P_r(r_k, \varphi'). \quad (6.56)$$

If we add and subtract

$$R(r_i, \varphi') \cos \theta(r_i, \varphi')(r_i \cos \varphi') + R(r_i, \varphi') \sin \theta(r_i, \varphi')(r_i \sin \varphi)$$

to the right-hand side of (6.56), we deduce that

$$P_{nrk}(r_{i+1}, \varphi) \geq P_{nri}(r_{i+1}, \varphi)$$

This is because

$$P_i(r_i, \varphi') \leq R(r_i, \varphi') \cos \theta(r_i, \varphi')(r_i \cos \varphi')$$
$$+ R(r_i, \varphi') \sin \theta(r_i, \varphi')(r_i \sin \varphi')$$
$$- R(r_k, \varphi') \cos \theta(r_k, \varphi')(r_i \cos \varphi')$$
$$- R(r_k, \varphi') \sin \theta(r_k, \varphi')(r_i \sin \varphi') + P_r(r_k, \varphi').$$

Inequality (6.52) and equality (6.54) prove that the radial path is the shortest one. Because payments are determined by shortest paths, all profit-maximizing payments $P(r_{i+1}, \varphi)$ are determined by radial paths. ∎

Theorem 6.5.4 *Optimal allocations are given by*

$$q_1(r_i, \varphi) = R(r_i) \cos \varphi,$$
$$q_2(r_i, \varphi) = R(r_i) \sin \varphi,$$

and payments $P(r_i, \varphi)$ do not depend on φ, that is,

$$P(r_i, \varphi) = P(r_i) = \sum_{l=1}^{i} r_l \left[R(r_l) - R(r_{l-1}) \right].$$

6.5 Multidimensional Types

Proof. Follows from Lemma 6.5.2 and Lemma 6.5.3. ∎

Theorem 6.5.4 allows us to solve Wilson's optimization problem in polar coordinates. To illustrate, suppose the uniform probability density over types. In the discrete case, the probability that a type is exactly within distance r_i from the origin is $f(r_i) = \frac{2i}{n(n+1)}$. Hence, the probability that a type is within distance r_i from the origin is $F(r_i) = \frac{i(i+1)}{n(n+1)}$. Problem $(OPT-W)$ is now reduced to a standard one-dimensional profit-maximization problem that can be successfully solved by following the approach of Section 6.2.2. Recall the general expression for the virtual valuation:

$$\mu(\mathcal{A}_i) = v(\mathcal{A}_i|i) - \frac{1-F(i)}{f_i}[v(\mathcal{A}_i|i+1) - v(\mathcal{A}_i|i)],$$

which in our case looks like

$$\mu(R(r_i)) = R(r_i)\frac{i}{n} - \frac{1-F(r_i)}{f(r_i)}\left[R(r_i)\frac{i+1}{n} - R(r_i)\frac{i}{n}\right],$$

$$\mu(R(r_i)) = R(r_i)\left(\frac{i}{n} - \frac{1-F(r_i)}{nf(r_i)}\right).$$

Hence, the profit-maximizing problem can be written as

$$\Pi = \max_{\{R(r_i)\}_{i=1}^n} \sum_{i=1}^n f(r_i)\left[R(r_i)\left(\frac{i}{n} - \frac{1-F(r_i)}{nf(r_i)}\right) - C(R(r_i))\right],$$

and can be solved type by type in the following formulation:

$$\max_{R(r_i)} R(r_i)\left(\frac{i}{n} - \frac{1-\frac{i(i+1)}{n(n+1)}}{\frac{2i}{(n+1)}}\right) - \frac{R(r_i)^2}{2}. \tag{OPT-W'}$$

Solving $(OPT-W)'$, we obtain the expression for the optimal choice of $R(r_i)$:

$$R(r_i) = \max\left(0, \frac{2i^2 + i(i+1) - n(n+1)}{2in}\right),$$

$$R(r_i) = \frac{n+1}{2i}\max\left(0, \frac{2i}{n+1}r_i + \frac{i+1}{n+1}r_i - 1\right). \tag{6.57}$$

The expression (6.57) is the discrete analog of Wilson's continuous solution to $(OPT-W)$ given by (6.38).

6.5.2 Capacity-Constrained Bidders

In our second example, the type of an agent is a pair of numbers (i, j). The first is her marginal value, the second her capacity.[23] Let the range of i be $R = \{1, \ldots, r\}$ and the range of j be $K = \{1, \ldots, k\}$. Let f_{ij} be the probability that an agent has type (i, j). Types are assumed to be independent. The value that an agent of type (i, j) assigns to q units will be written $v(q|i, j) = i \min\{q, j\}$. Observe that $v(q|i, j)$ satisfies increasing differences. That is, if $q' \geq q$ and $(i', j') \geq (i, j)$, then

$$v(q'|i', j') - v(q|i', j') \geq v(q'|i, j) - v(q|i, j).$$

We will identify the revenue-maximizing auction for selling Q units of an identical good to such capacity-constrained agents. Without loss of generality, we can assume that the amount assigned to an agent who reports type (i, j) will be at most j. As in the single-object case, we restrict attention to allocation rules that are anonymous. Hence, we focus on agent 1 only. Denote by $\mathcal{A}_{ij}$ is the expected amount assigned to agent 1 when she reports (i, j). Because agent 1 receives at most j in any allocation, her expected payoff is $i\mathcal{A}_{ij} = v(\mathcal{A}_{i,j}|i, j)$. Let P_{ij} be her expected payment if she reports (i, j). Next, we discuss the relevant incentive compatibility constraints and identify the ones that are redundant.

The BNIC Constraints

There are eight types of BNIC constraints:

(1) Horizontal Upward BNIC (HUBNIC)
Type (i, j) reporting (i', j) where $i' > i$:

$$v(\mathcal{A}_{i,j}|i, j) - P_{i,j} \geq v(\mathcal{A}_{i',j}|i, j) - P_{i',j}.$$

(2) Horizontal Downward BNIC (HDBNIC)
Type (i, j) reporting (i', j) where $i' < i$:

$$v(\mathcal{A}_{i,j}|i, j) - P_{i,j} \geq v(\mathcal{A}_{i',j}|i, j) - P_{i',j}.$$

(3) Vertical Upward BNIC (VUBNIC)
Type (i, j) reporting (i, j') where $j' > j$:

$$v(\mathcal{A}_{i,j}|i, j) - P_{i,j} \geq v(\mathcal{A}_{i,j'}|i, j) - P_{i,j'}.$$

(4) Vertical Downward BNIC (VDBNIC)
Type (i, j) reporting (i, j') where $j' < j$:

$$v(\mathcal{A}_{i,j}|i, j) - P_{i,j} \geq v(\mathcal{A}_{i,j'}|i, j) - P_{i,j'}.$$

---

[23] This section is based on Malakhov and Vohra (2007). The continuous type version is examined in Iyenagar and Kumar (2008).

6.5 Multidimensional Types

(5) Diagonal Downward BNIC (DDBNIC)
Type (i, j) reporting (i', j') where $i' > i$ and $j' < j$:
$$v(\mathcal{A}_{i,j}|i, j) - P_{i,j} \geq v(\mathcal{A}_{i',j'}|i, j) - P_{i',j'}.$$

(6) Diagonal Upward BNIC (DUBNIC)
Type (i, j) reporting (i', j') where $i' < i$ and $j' > j$:
$$v(\mathcal{A}_{i,j}|i, j) - P_{i,j} \geq v(\mathcal{A}_{i',j'}|i, j) - P_{i',j'}.$$

(7) Leading Diagonal Downward BNIC (LDDBNIC)
Type (i, j) reporting (i', j') where $i' < i$ and $j' < j$:
$$v(\mathcal{A}_{i,j}|i, j) - P_{i,j} \geq v(\mathcal{A}_{i',j'}|i, j) - P_{i',j'}.$$

(8) Leading Diagonal Upward BNIC (LDUBNIC)
Type (i, j) reporting (i', j') where $i' > i$ and $j' > j$:
$$v(\mathcal{A}_{i,j}|i, j) - P_{i,j} \geq v(\mathcal{A}_{i',j'}|i, j) - P_{i',j'}.$$

The new wrinkle that multidimensionality adds are the diagonal BNIC constraints. One might wish they are redundant. In general, they are not.

Simplification of Incentive Constraints

We now impose the assumption that no agent is able to inflate their capacity. A justification for such an assumption is that an agent who "bids" more than his capacity and is then called on to consume it suffers a large penalty. In effect we rule out free disposal.[24] Under the no-inflation assumption, the VUBNIC, DUBNIC, and LDUBNIC constraints can be thrown out.

Theorem 6.5.5 *An allocation rule satisfies HUBNIC and HDBNIC if it is monotonic in the 'i' component. That is, for all $i \geq i'$, we have $\mathcal{A}_{ij} \geq \mathcal{A}_{i',j}$.*

The proof is standard and omitted. The absence of VUBNIC means that a similar monotonicity result does not hold for the 'j' argument. In other words, it is not true that $\mathcal{A}_{ij} \geq \mathcal{A}_{ij'}$ when $j \geq j'$. This is because both upward and downward incentive constraints are needed to ensure monotonicity of an allocation rule. However, we will assume monotonicity in both components of the allocation rule.

We now show that the two remaining diagonal BNIC constraints are redundant.

Theorem 6.5.6 *The LDDBNIC constraints relating type (i', j') to (i, j) where $(i', j') > (i, j)$,*

$$v(\mathcal{A}_{i',j'}|i', j') - P_{i',j'} \geq v(\mathcal{A}_{i,j}|i', j') - P_{i,j}, \qquad (6.58)$$

are implied by HDBNIC, VDBNIC, and monotonicity of the allocation rule.

---

[24] If we flip to a procurement setting, the no-inflation assumption would mean that a bidder who bids to supply more than they can will suffer a large penalty.

Proof. To see that (6.58) is implied by HDBNIC and VDBNIC, consider

$$v(\mathcal{A}_{i',j'}|i', j') - P_{i',j'} \geq v(\mathcal{A}_{i,j'}|i', j') - P_{i,j'}$$

and

$$v(\mathcal{A}_{i,j'}|i, j') - P_{i,j'} \geq v(\mathcal{A}_{i,j}|i, j') - P_{i,j}.$$

Adding these two inequalities together yields:

$$v(\mathcal{A}_{i',j'}|i', j') - P_{i',j'} + v(\mathcal{A}_{i,j'}|i, j') \geq v(\mathcal{A}_{i,j'}|i', j') + v(\mathcal{A}_{i,j}|i, j') - P_{i,j}.$$

Rearranging:

$$v(\mathcal{A}_{i',j'}|i', j') - v(\mathcal{A}_{i,j}|i', j') - [P_{i',j'} - P_{i,j}] \geq$$

$$\geq [v(\mathcal{A}_{i,j'}|i', j') - v(\mathcal{A}_{i,j}|i', j')] - [v(\mathcal{A}_{i,j'}|i, j') - v(\mathcal{A}_{i,j}|i, j')].$$

The increasing differences property and the monotonicity of the allocation imply that

$$[v(\mathcal{A}_{i,j'}|i', j') - v(\mathcal{A}_{i,j}|i', j')] - [v(\mathcal{A}_{i,j'}|i, j') - v(\mathcal{A}_{i,j}|i, j')] \geq 0,$$

and hence we have that

$$v(\mathcal{A}_{i',j'}|i', j') - v(\mathcal{A}_{i,j}|i', j') \geq P_{i',j'} - P_{i,j},$$

which is equivalent to (6.58) and proves the claim. ∎

Theorem 6.5.7 *Under the assumption that no agent can inflate their capacity, the DDBNIC constraints are redundant.*

Proof. Let $i' \geq i$ and $j' \geq j$. Consider the DDBNIC constraint

$$v(\mathcal{A}_{i,j'}|i, j') - P_{i,j'} \geq v(\mathcal{A}_{i',j}|i, j') - P_{i',j}.$$

When we substitute in our expression for v, we obtain:

$$i\mathcal{A}_{i,j'} - P_{i,j'} \geq i\mathcal{A}_{i',j} - P_{i',j}. \tag{6.59}$$

We show that it is implied by the addition of the following VDBNIC and HUBNIC constraints:

$$v(\mathcal{A}_{i,j'}|i, j') - P_{i,j'} \geq v(\mathcal{A}_{i,j}|i, j') - P_{ij}, \tag{6.60}$$

$$v(\mathcal{A}_{i,j}|i, j) - P_{ij} \geq v(\mathcal{A}_{i',j}|i, j) - P_{i',j}. \tag{6.61}$$

Adding (6.60) and (6.61) yields:

$$v(\mathcal{A}_{i,j'}|i, j') - P_{i,j'} + v(\mathcal{A}_{i,j}|i, j) \geq v(\mathcal{A}_{i,j}|i, j')$$
$$+ v(\mathcal{A}_{i',j}|i, j) - P_{i',j}. \tag{6.62}$$

Substitute in our expression for v:

$$i\mathcal{A}_{i,j'} - P_{i,j'} + i\mathcal{A}_{i,j} \geq i\mathcal{A}_{i,j} + i\mathcal{A}_{i',j} - P_{i',j}.$$

Canceling common terms yields (6.59). ∎

6.5 Multidimensional Types

Theorem 6.5.8 *Only the adjacent downward constraints w.r.t. $(i + 1, j)$ and (i, j), and w.r.t $(i, j + 1)$, and (i, j) matter out of all horizontal and vertical downward constraints. Only the adjacent upward constraints w.r.t. (i, j) and $(i + 1, j)$, and w.r.t. (i, j) and $(i, j + 1)$ matter out of all horizontal and vertical upward constraints.*

The proof is standard and omitted.

Theorem 6.5.9 *If an adjacent HDBNIC or VDBNIC constraint binds, the corresponding adjacent upward BNIC constraint are satisfied.*

Proof. Given that the corresponding downward adjacent BNIC constraint binds, then the upward BNIC constraints are satisfied. Indeed, if

$$v(\mathcal{A}_{i+1,j}|i+1, j) - v(\mathcal{A}_{i,j}|i+1, j) = P_{i+1,j} - P_{i,j},$$

then by increasing differences and monotonicity of the allocation rule:

$$v(\mathcal{A}_{i+1,j}|i, j) - v(\mathcal{A}_{i,j}|i, j) \leq P_{i+1,j} - P_{i,j},$$

which is the corresponding upward constraint

$$v(\mathcal{A}_{i,j}|i, j) - P_{i,j} \geq v(\mathcal{A}_{i+1,j}|i, j) - P_{i+1,j}.$$

So the corresponding adjacent upward BNIC constraint is satisfied. The argument is exactly the same w.r.t. the j dimension. ∎

When an adjacent HDBNIC constraint does not bind, the corresponding adjacent HUBNIC constraint is not automatically satisfied. Also some of the adjacent downward BNIC constraints will be slack because not all arcs are likely to be used in a shortest-path tree.

Summarizing, the only BNIC constraints that matter are the adjacent HUBNIC, HDBNIC and VDBNIC.

Optimal Auction Formulation and Solution

Denote by $\alpha_{ij}[\mathbf{t}]$ the *actual* allocation that an agent with type (i, j) will receive under allocation rule $\mathcal{A}$ when the announced profile is $\mathbf{t}$. In the case when type (i, j) does not appear in the profile $\mathbf{t}$, we take $\alpha_{ij}[\mathbf{t}] = 0$. We will have cause to study how the allocation for agent 1 with a type, (i, j), say, will change when the types of the other $n - 1$ agents change. In these cases, we will write $\alpha_{ij}(\mathbf{t})$ as $\alpha_{ij}[(i, j), \mathbf{t}^{-1}]$. Then $\mathcal{A}_{ij} = \sum_{\mathbf{t}^{-1}} \pi(\mathbf{t}^{-1}) \alpha_{ij}[(i, j), \mathbf{t}^{-1}]$ where $\pi(\mathbf{t}^{-1})$ is the probability of profile $\mathbf{t}^{-1}$ being realized and the sum is over all possible profiles. Let $n_{ij}(\mathbf{t})$ denote the number of bidders in the profile $\mathbf{t}$ with type (i, j).

We now formulate the problem of finding the revenue-maximizing monotone mechanism as a linear program in the following way. We omit all diagonal BNIC constraints as well as all the upward vertical and horizontal BNIC constraints. We identify an optimal solution to this relaxation in which the adjacent HDBNIC constraints bind. Theorem 6.5.9 will ensure that HUBNIC will be satisfied. The problem we study is (*OPT*):

$$Z = \max_{P_{ij}} n \sum_{i \in R} \sum_{j \in K} f_{ij} P_{ij}$$

s.t. $v(\mathcal{A}_{ij}|i, j) - P_{ij} \geq v(\mathcal{A}_{i-1,j}|i, j) - P_{i-1,j} \quad \forall i \in R, j \in K$

$v(\mathcal{A}_{ij}|i, j) - P_{ij} \geq v(\mathcal{A}_{i,j-1}|i, j) - P_{i,j-1} \quad \forall i \in R, j \in K$

$\mathcal{A}_{ij} \geq \mathcal{A}_{i'j'} \quad \forall (i, j) \geq (i', j')$

$\mathcal{A}_{ij} = \sum_{\mathbf{t}^{-1}} \pi(\mathbf{t}^{-1}) \alpha_{ij}[(i, j), \mathbf{t}^{-1}] \quad \forall i \in R, j \in K$

$\sum_{i \in R} \sum_{j \in K} n_{ij}(\mathbf{t}) \alpha_{ij}[\mathbf{t}] \leq Q \quad \forall \mathbf{t} \in T$

$\alpha_{ij}[\mathbf{t}] \leq j \quad \forall i \in R, j \in K \ \forall \mathbf{t} \in T$

To describe the network representation of this linear program, fix the $\mathcal{A}_{ij}$'s. For each type (i, j), introduce a node including the dummy type $(0, 0)$. For each pair $(i, j), (i + 1, j)$, introduce a directed arc from (i, j) to $(i + 1, j)$, with length $v(\mathcal{A}_{i+1,j}|i + 1, j) - v(\mathcal{A}_{ij}|i + 1, j) = (i + 1)\mathcal{A}_{i+1,j} - (i + 1)\mathcal{A}_{ij}$. Similarly, introduce a directed arc from (i, j) to $(i, j + 1)$ of length $i\mathcal{A}_{i,j+1} - i\mathcal{A}_{ij}$. Then P_{ij} will be the length of the shortest path from the dummy type $(0, 0)$ to (i, j). We show that the shortest path from $(1, 1)$ to (i, j) is $(1, 1) \to (1, 2) \to (1, 3) \cdots \to (1, j) \to (2, j) \cdots \to (i, j)$. An example of such paths is provided in Figure 6.3 below:

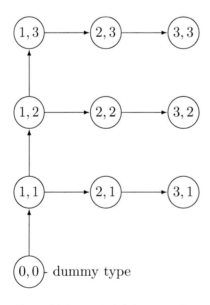

Figure 6.3 Examples of shortest path.

6.5 Multidimensional Types

Theorem 6.5.10

$$P_{ij} = v(\mathcal{A}_{ij}|i, j) - \sum_{r=1}^{i-1}[v(\mathcal{A}_{rj}|r + 1, j) - v(\mathcal{A}_{rj}|r, j)]$$

$$= \sum_{r=1}^{i} r(\mathcal{A}_{rj} - \mathcal{A}_{r-1,j}) - \mathcal{A}_{11} + P_{11}.$$

Proof. It suffices to show that the shortest path from $(1, 1)$ to (i, j) is straight up and across. The proof is by induction. It is clearly true for nodes $(1, 2)$ and $(2, 1)$. Consider the node $(2, 2)$. The length of $(1, 1) \to (2, 1) \to (2, 2)$ is

$$2\mathcal{A}_{22} - 2\mathcal{A}_{21} + 2\mathcal{A}_{21} - 2\mathcal{A}_{11} + P_{11} = 2\mathcal{A}_{22} - 2\mathcal{A}_{11} + P_{11}.$$

The length of the path $(1, 1) \to (1, 2) \to (2, 2)$ is

$$2\mathcal{A}_{22} - 2\mathcal{A}_{1,2} + \mathcal{A}_{1,2} - \mathcal{A}_{11} + P_{11} = 2\mathcal{A}_{22} - \mathcal{A}_{12} - \mathcal{A}_{11} + P_{11}.$$

The difference in length between the first and the second path is

$$(2\mathcal{A}_{22} - 2\mathcal{A}_{11}) - (2\mathcal{A}_{22} - \mathcal{A}_{12} - \mathcal{A}_{11}) = \mathcal{A}_{12} - \mathcal{A}_{11} \geq 0,$$

where the last inequality follows by monotonicity of the $\mathcal{A}$'s.

Now suppose the claim is true for all nodes (i, j) where $i, j \leq q - 1$. The shortest path from $(1, 1)$ to $(1, q)$ is clearly up the top. A similar argument to the previous one shows that the shortest path from $(1, 1)$ to $(2, q)$ is also up the top and across. Consider now node $(3, q)$. There are two candidates for a shortest path from $(1, 1)$ to $(3, q)$. One is $(1, 1) \to (1, q - 1) \to (3, q - 1) \to (3, q)$. This path has length

$$3\mathcal{A}_{3q} - 3\mathcal{A}_{3,q-1} + 3\mathcal{A}_{3,q-1} - 3\mathcal{A}_{2,q-1} + 2\mathcal{A}_{2,q-1} - 2\mathcal{A}_{1,q-1} + P_{1,q-1}$$

$$= 3\mathcal{A}_{3q} - \mathcal{A}_{2,q-1} - 2\mathcal{A}_{1,q-1} + P_{1,q-1}.$$

The other path, $(1, 1) \to (1, q) \to (3, q)$, has length

$$3\mathcal{A}_{3q} - 3\mathcal{A}_{2q} + 2\mathcal{A}_{2q} - 2\mathcal{A}_{1,q} + \mathcal{A}_{1,q} - \mathcal{A}_{1,q-1} + P_{1,q-1}$$

$$= 3\mathcal{A}_{3q} - \mathcal{A}_{2q} - \mathcal{A}_{1,q} - \mathcal{A}_{1,q-1} + P_{1,q-1}.$$

The difference in length between the first and second path is

$$\mathcal{A}_{2q} + \mathcal{A}_{1q} + \mathcal{A}_{1,q-1} - \mathcal{A}_{2,q-1} - 2\mathcal{A}_{1,q-1}$$

$$= \mathcal{A}_{2q} + \mathcal{A}_{1q} - \mathcal{A}_{2,q-1} - \mathcal{A}_{1,q-1} \geq 0.$$

Again the last inequality follows by monotonicity of the $\mathcal{A}$'s.

Proceeding inductively in this way, we can establish the claim for nodes of the form (i, q) where $i \leq q - 1$ and for (q, j) where $j \leq q - 1$. It remains then to prove the claim for node (n, q). One path is $(1, q - 1) \to (q, q - 1) \to$

(q, q) and has length

$$qA_{qq} - qA_{q,q-1} + qA_{q,q-1} - qA_{q-1,q-1}$$
$$+ (q-1)A_{q-1,q-1} - (q-1)A_{q-2,q-1} + \cdots + P_{1,q-1}$$
$$= qA_{qq} - A_{q-1,q-1} - A_{q-2,q-1} - \cdots + P_{1,q-1}.$$

The length of the other path, $(1, 1) \to (1, q) \to (q, q)$, is

$$qA_{qq} - qA_{q-1,q} + (q-1)A_{q-1,q} - (q-1)A_{q-2,q}$$
$$+ \cdots + A_{1q} - A_{1,q-1} + P_{1,q-1}.$$

Again, by the monotonicity of the $\mathcal{A}$'s, the second path is shorter than the first. ∎

Substituting the expression for P_{ij} derived into the objective function for problem (*OPT*), we get

$$\sum_{j=1}^{k} \sum_{i=1}^{r} \{f_{ij} v(\mathcal{A}_{ij}|i, j) + (1 - F_j(i))[v(\mathcal{A}_{ij}|i, j) - v(\mathcal{A}_{ij}|i+1, j)]\},$$

where $F_j(i) = \sum_{t=1}^{i} f_{tj}$. The expression within the summation term can be rewritten as

$$f_{ij}\{v(\mathcal{A}_{ij}|i, j) + \frac{1 - F_j(i)}{f_{ij}}[v(\mathcal{A}_{ij}|i, j) - v(\mathcal{A}_{ij}|i+1, j)]\}.$$

The particular functional form of v allows us to rewrite this as:

$$f_{ij}\mathcal{A}_{ij}\left(i - \frac{1 - F_j(i)}{f_{ij}}\right).$$

We can think of the term $(i - \frac{1-F_j(i)}{f_{ij}})$ as the virtual valuation conditional on wanting to consume at most j units. Problem (*OPT*) becomes:

$$Z = \max_{\{\alpha, \mathcal{A}\}} n \sum_{i \in R} \sum_{j \in K} f_{ij}\mathcal{A}_{ij}\left(i - \frac{1 - F_j(i)}{f_{ij}}\right)$$

$$\text{s.t. } \mathcal{A}_{ij} \geq \mathcal{A}_{i'j'} \; \forall (i, j) \geq (i', j')$$

$$\mathcal{A}_{ij} = \sum_{\mathbf{t}^{-1}} \pi(\mathbf{t}^{-1})\alpha_{ij}[(i, j), \mathbf{t}^{-1}] \; \forall (i, j)$$

$$\sum_{i \in R} \sum_{j \in K} n_{ij}(\mathbf{t})\alpha_{ij}[\mathbf{t}] \leq Q \; \forall \mathbf{t}$$

$$\alpha_{ij}[\mathbf{t}] \leq j \; \forall i, j \; \forall \mathbf{t}.$$

6.5 Multidimensional Types

Substituting out the $\mathcal{A}_{ij}$ variables yields:

$$Z = \max_{\{\alpha, \mathcal{A}\}} n \sum_{i \in R} \sum_{j \in K} f_{ij} \sum_{\mathbf{t}^{-1}} \pi(\mathbf{t}^{-1}) \alpha_{ij}[(i,j), \mathbf{t}^{-1}] \left(i - \frac{1 - F_j(i)}{f_{ij}} \right)$$

s.t. $\sum_{\mathbf{t}^{-1}} \pi(\mathbf{t}^{-1}) \alpha_{ij}[(i,j), \mathbf{t}^{-1}] \geq \sum_{\mathbf{t}^{-1}} \pi(\mathbf{t}^{-1}) \alpha_{i'j'}[(i',j'), \mathbf{t}^{-i}] \; \forall (i,j) \geq (i',j')$

$$\sum_{i \in R} \sum_{j \in K} n_{ij}(\mathbf{t}) \alpha_{ij}[\mathbf{t}] \leq Q \; \forall \mathbf{t}$$

$$\alpha_{ij}[\mathbf{t}] \leq j \; \forall i, j \; \forall \mathbf{t}.$$

Theorem 6.5.11 *Suppose the conditional virtual values are monotone, that is,*

$$i - \frac{1 - F_j(i)}{f_{ij}} \geq i' - \frac{1 - F_{j'}(i')}{f_{i'j'}} \; \forall \; (i,j) \geq (i',j').$$

Then the following procedure describes an optimal solution to problem (OPT). Select the pair (i,j) that maximizes $i - \frac{1 - F_j(i)}{f_{ij}}$ and increase α_{ij} until it reaches its upper bound or the supply is exhausted, whichever comes first. If the supply is not exhausted, repeat.

Proof. If we ignore the monotonicity condition

$$\sum_{\mathbf{t}^{-1}} \pi(\mathbf{t}^{-1}) \alpha_{ij}[(i,j), \mathbf{t}^{-1}] \geq \sum_{\mathbf{t}^{-1}} \pi(\mathbf{t}^{-1}) \alpha_{i'j'}[(i',j'), \mathbf{t}^{-1}],$$

the optimization problem reduces to a collection of optimization problems one for each profile $\mathbf{t}$:

$$Z(t) = \max_{\{\alpha\}} \sum_{i} \sum_{j} n_{ij}(\mathbf{t}) \alpha_{ij}[\mathbf{t}] \left(i - \frac{1 - F_j(i)}{f_{ij}} \right)$$

s.t. $\sum_{i \in R} \sum_{j \in K} n_{ij}(\mathbf{t}) \alpha_{ij}[\mathbf{t}] \leq Q$

$$\alpha_{ij}[\mathbf{t}] \leq j \; \forall i, j \; \forall \mathbf{t}$$

This is an instance of a continuous knapsack problem with upper bound constraints on the variables that can be solved in the usual greedy manner. In each profile, allocate as much as possible to the agents with highest conditional virtual values. Once they are saturated, proceed to the next highest, and so on. Monotonicity of the conditional virtual values ensures that the resulting solution satisfies the omitted monotonicity constraint. ∎

Requiring the conditional virtual values to be monotone is clearly a more demanding requirement than the analogous requirement when types are one-dimensional. If $\frac{1 - F_j(i)}{f_{ij}}$ is nonincreasing in (i, j), then the conditional virtual values are monotone. This condition can be interpreted as a type of affiliation. One example of a distribution on types that yields monotone conditional virtual values is when each component of the type is independent of the other and the i component is drawn from a distribution that satisfies the monotone hazard condition while the j component is drawn from a uniform distribution. When the conditional virtual values are not monotone, one can solve the problem using "ironed" conditional virtual values. The details are omitted. It is easy to see that the solution to the continuous case is exactly as described in Theorem 6.5.11.

Monotonicity and the Conditional Virtual Values

The theorems above identify the optimal mechanism from among a restricted class of mechanisms, one where the associated allocation rule is monotone. In this section we drop this restriction. When the conditional virtual values are monotone, we show that there is an optimal mechanism whose associated allocation rule is monotone.

The argument is in two steps. In the first step, we upper-bound the value of Z in (OPT) by using a relaxation.

Lemma 6.5.12 *Fix an allocation rule $\mathcal{A}$. The expected revenue from the allocation rule $\mathcal{A}$ is bounded above by*

$$\sum_{i \in R} \sum_{j \in K} f_{ij} \left(i - \frac{1 - F_j(i)}{f_{ij}} \right) \mathcal{A}_{ij}.$$

Proof. Consider the following relaxation of (OPT), called $(rOPT)$, where the choice of $\mathcal{A}$ is fixed:

$$Z(\mathcal{A}) = \max_{P_{ij}} n \sum_{i \in R} \sum_{j \in K} f_{ij} P_{ij}$$

$$\text{s.t. } v(\mathcal{A}_{ij}|i, j) - P_{ij} \geq v(\mathcal{A}_{i-1,j}|i, j) - P_{i-1,j} \qquad (6.63)$$

$$\mathcal{A}_{ij} = \sum_{\mathbf{t}^{-1}} \pi(\mathbf{t}^{-1}) \alpha_{ij}[(i, j), \mathbf{t}^{-1}] \; \forall (i, j)$$

$$\sum_{i \in R} \sum_{j \in K} n_{ij}(\mathbf{t}) \alpha_{ij}[\mathbf{t}] \leq Q \; \forall \mathbf{t}$$

$$\alpha_{ij}[\mathbf{t}] \leq j \; \forall i, j \; \forall \mathbf{t}$$

To prove the lemma, it suffices to show that

$$Z(\mathcal{A}) = \sum_{i \in R} \sum_{j \in K} f_{ij} \left(i - \frac{1 - F_j(i)}{f_{ij}} \right) \mathcal{A}_{ij}.$$

We do so by induction.

Fix a j and let i be the smallest number such that $\mathcal{A}_{ij} > 0$. Clearly, in an optimal solution to $(rOPT)$, $P_{ij} = i\mathcal{A}_{ij}$. The induction hypothesis is that $P_{i+k,j} = (i+k)\mathcal{A}_{i+k,j} - \sum_{s=1}^{i+k-1} \mathcal{A}_{sj}$.

6.5 Multidimensional Types

In an optimal solution to $(rOPT)$, the constraint (6.63) linking type $(i+1, j)$ to (i, j) must bind. Otherwise, increase the value of $P_{i+1,\cdot}$. Therefore,

$$(i+k+1)\mathcal{A}_{i+k+1,j} - P_{i+k+1,j} = (i+k+1)\mathcal{A}_{i+k,j} - P_{i+k,j} = \sum_{s=1}^{i+k} \mathcal{A}_{sj}.$$

To complete the proof, we mimic the calculation immediately following the proof of Theorem 6.5.10. ∎

It is now easy to see that when the conditional virtual values are monotone, the mechanism of Theorem 6.5.11 achieves the upper bound implied by Lemma 6.5.12 and so must be optimal.

CHAPTER 7

Rationalizability

This chapter is quite brief. Its subject has fallen out of fashion. Nonetheless, it is worth our consideration because, as noted in Rochet (1987), we can think of the rationalizability problem as being "dual" to the problem of characterizing IDS. The basic question is this: When is a sequence of observed purchase decisions consistent with the purchaser maximizing a concave utility function $u(\cdot)$? To see the connection to mechanism design, suppose a sequence of purchase decisions (p_i, x_i), $i = 1, \ldots, n$, where $p_i \in \mathbb{R}_+^m$ and $x_i \in \mathbb{R}_+^m$ are price and purchased quantity vectors, respectively. Via the taxation principle, we can think of this sequence of price-quantity pairs as a menu offered by a mechanism. Thus, the mechanism is given and we ask for what types or preferences is this mechanism incentive compatible. Given the close connection, one should expect that the same machinery could be deployed, and indeed this is the case. We confine ourselves to two examples. In the first, we assume the utility function of the purchaser to be quasilinear.

7.1 THE QUASILINEAR CASE

A sequence of purchase decisions (p_i, x_i), $i = 1, \ldots, n$, is **rationalizable** by a concave quasilinear utility function $u : \mathbb{R}_+^m \mapsto \mathbb{R}$ if for some budget B and for all i

$$x_i \in \arg\max\{u(x) + s_i : p_i \cdot x + s_i = B, x \in \mathbb{R}_+^m, s_i \geq 0\}.$$

To determine if a sequence is rationalizable, we suppose it is and ask what conditions it must satisfy, and then check to see if those conditions are indeed sufficient.

If the sequence $\{(p_i, x_i)\}_{i=1}^n$ is rationalizable, it must be the case that at price p_i, if $p_i \cdot x_j \leq B$, it must be that x_j delivers less utility than x_i. Hence,

$$u(x_i) + B - p_i \cdot x_i \geq u(x_j) + B - p_i \cdot x_j$$
$$\Rightarrow u(x_j) - u(x_i) \leq p_i \cdot (x_j - x_i).$$

Thus, given the sequence $\{(p_i, x_i)\}_{i=1}^n$, we formulate the system:

$$y_j - y_i \leq p_i \cdot (x_j - x_i), \quad \forall i, j \text{ s.t. } p_i \cdot x_j \leq B \tag{7.1}$$

If this system is feasible, we can use any feasible choice of $\{y_j\}_{j=1}^n$ to construct a concave utility function that rationalizes the sequence $\{p_i, x_i\}_{i=1}^n$. Determining feasibility of (7.1) is straightforward. We associate a network with (7.1) in the usual way: One node for each i and for each ordered pair (i, j) such that $p_i \cdot x_j \leq B$, an arc with length $p_i \cdot (x_j - x_i)$. The system (7.1) is feasible if the associated network has no negative-length cycles. For example, if $i_1 \to i_2 \to i_3 \to \cdots \to i_k \to i_1$ is a cycle, then

$$p_{i_1} \cdot (x_{i_2} - x_{i_1}) + \cdots + p_{i_k} \cdot (x_{i_1} - x_{i_k}) \geq 0.$$

Suppose our associated network has no negative-length cycles. Choose any vector y that satisfies (7.1) and set $u(x_i) = y_i$. For any other $x \in \mathbb{R}_+^n$, set

$$u(x) = \min_{i=1,\ldots,n} \{u(x_i) + p_i \cdot (x - x_i)\}.$$

Because $u(x)$ is the minimum of a collection of linear functions with positive slope, it is easy to see that u is concave. It remains to verify that for any p_r and x such that $p_r \cdot x \leq B$, that $u(x_r) + s_r \geq u(x) + B - p_r \cdot x$. From the definition of u, we know that $u(x) \leq u(x_r) + p_r \cdot (x - x_r) \leq u(x_r)$. Adding B to both sides of the last inequality achieves the desired verification.

It is also easy to see from the argument that the absence of negative-length cycles is a necessary and sufficient condition for the sequence to be rationalizable by a concave quasilinear utility function.

7.2 THE GENERAL CASE

Here we consider the case when utilities are not quasilinear. This was first considered in Afriat (1967). A sequence of purchase decisions $\{p_i, x_i\}_{i=1}^n$ is rationalizable by a locally nonsatiated concave utility function $u : \mathbb{R}_+^m \mapsto \mathbb{R}$ if for some budget B and for all i,

$$x_i \in \arg\max\{u(x) : p_i \cdot x \leq B, x \in \mathbb{R}_+^m\}.$$

A utility function u is **locally nonsatiated** if for every x, and in every neighborhood of x, there is a y such that $u(y) > u(x)$.

As before, we identify conditions that a rationalizable sequence must satisfy. Suppose the sequence $\{p_i, x_i\}_{i=1}^n$ is indeed rationalizable. Because

$$x_i \in \arg\max\{u(x) : p_i \cdot x \leq B\},$$

each x_i must satisfy the usual first-order conditions of optimality. In particular, if $s_i > 0$ is the optimal Lagrange multiplier, it must be that

$$x_i \in \arg\max\{u(x) + s_i(B - p_i \cdot x)\}.$$

Note that local insatiability forces each $s_i > 0$. Hence, for all $j \neq i$,

$$u(x_j) + s_i(B - p_i \cdot x_j) \leq u(x_i) + s_i(B - p_i \cdot x_i)$$
$$\Rightarrow u(x_j) \leq u(x_i) + s_i p_i \cdot (x_j - x_i).$$

If we replace each $u(x_i)$ by y_i and set $a_{ij} = p_i \cdot (x_j - x_i)$, we obtain the system $\ell(A)$:

$$y_j \leq y_i + s_i a_{ij} \quad \forall i \neq j, \ 1 \leq i, j \leq n$$
$$s_i > 0 \ \forall 1 \leq i \leq n$$

Given a feasible solution to $\ell(A)$, we construct a locally nonsatiated concave utility function $u(\cdot)$ consistent with the sequence of purchase decisions (p_i, x_i) by setting:

$$u(x) = \min_{i=1,\ldots,n} \{y_i + s_i p_i \cdot (x - x_i)\}.$$

Hence, the sequence $\{p_i, x_i\}_{i=1}^n$ is rationalizable if and only if the system $\ell(A)$ is feasible. By itself, this is not very revealing. One would like an interpretation of the condition that $\ell(A)$ is feasible. This could be obtained by examining the Faraks alternative for this system. This is essentially what we will do, but not explicitly so.

If $p_i \cdot (x_j - x_i) \leq 0$, the utility function u must satisfy $u(x_j) \leq u(x_i)$, otherwise, with purchase price of p_i, bundle x_j costs less but provides higher utility. In other words, the bundle x_i is revealed preferred to x_j. Suppose now we have a sequence of decisions $(p_i, x_i), (p_j, x_j), (p_k, x_k), \ldots, (p_r, x_r)$, with

$$p_i \cdot (x_j - x_i) \leq 0, \quad p_j \cdot (x_k - x_j) \leq 0, \ldots, \quad p_r \cdot (x_i - x_r) \leq 0.$$

This means

$$u(x_i) \geq u(x_j) \geq u(x_k) \geq \cdots \geq u(x_r) \geq u(x_i).$$

Hence, $u(x_i) = u(x_j) = \cdots = u(x_r)$, and

$$p_i \cdot (x_j - x_i) = 0, \quad p_j \cdot (x_k - x_j) = 0, \ldots, \quad p_r \cdot (x_i - x_r) = 0.$$

The above *necessary* condition for rationalizability is sometimes called the **generalized axiom of revealed preference** or GARP. It can be described in graph theoretic terms as follows. Let A be a $n \times n$ matrix where $a_{ij} = p_i \cdot (x_j - x_i)$. Associate with the matrix A a directed graph $D(A)$ as follows: introduce a vertex for each index, and for each ordered pair (i, j) an edge with length a_{ij}. The matrix A is said to satisfy GARP if every negative length cycle in $D(A)$ contains at least one edge of positive weight.

We now state Afriat's Theorem:

7.2 The General Case

Theorem 7.2.1 $\ell(A)$ *is feasible iff.* $D(A)$ *satisfies GARP.*

A number of proofs of the theorem exist.[1] Here we give a proof that makes explicit the network structure inherent in $\ell(A)$.[2]

To each $s \in \mathbb{R}_+^m$ and matrix A with zeros on the diagonals, we associate a directed graph $D(A, s)$ as follows: introduce a vertex for each index, and for each ordered pair (i, j) an edge with length $s_i a_{ij}$. Notice that $D(A) = D(A, e)$ where e is the n-vector of all 1's.

Now fix $s \in \mathbb{R}_+^m$. Then feasibility of $\ell(A)$ reduces to identifying $y \in \mathbb{R}^m$ such that $y_j - y_i \leq s_i a_{ij}$ for all $i \neq j$. This system is feasible if $D(A, s)$ contains no negative cycles. Assuming feasibility, we can choose the y's as follows: set $y_1 = 0$ and y_j to be the length of the shortest path from 1 to i in $D(A, s)$. Afriat's Theorem can be rephrased as:

Theorem 7.2.2 *There is an* $s \in \mathbb{R}_+^m$ *such that* $D(A, s)$ *contains no negative cycles iff.* $D(A, e)$ *satisfies GARP.*

Proof. If there is an $s \in \mathbb{R}_+^m$ such that $D(A, s)$ contains no negative cycles, the system of inequalities $\ell(A)$ (with s fixed) is feasible. So we can construct a utility function $u(\cdot)$ consistent with the sequence of purchase decisions. Therefore, $D(A, e)$ must satisfy GARP. We next prove the nontrivial direction. Suppose $D(A, e)$ satisfies GARP. We prove there exists $s \in \mathbb{R}_+^m$ such that $D(A, s)$ has no negative cycles.

Let $S = \{(i, j) : a_{ij} < 0\}$, $Z = \{(i, j) : a_{ij} = 0\}$, and $T = \{(i, j) : a_{ij} > 0\}$. Consider the weighted digraph G with edges in $S \cup Z$, where arcs $(i, j) \in S$ are given weight $w_{ij} = -1$, and arcs $(i, j) \in Z$ are given weight $w_{ij} = 0$. Because $D(A, e)$ satisfies GARP, G does not contain a negative-length cycle. Hence, there exists a set of potentials $\{\phi_j\}$ on the nodes such that

$$\phi_j \leq \phi_i + a_{ij}, \quad \forall \ (i, j) \in S \cup Z.$$

Without loss of generality, we relabel the vertices so that $\phi_n \leq \phi_{n-1} \leq \cdots \leq \phi_1$. Choose $\{s_i\}$ nondecreasing so that

$$s_i \times \min_{(i,j) \in T} a_{ij} \geq (n-1) \times s_{i-1} \max_{(i,j) \in S} (-a_{ij}) \text{ if } \phi_i < \phi_{i-1},$$

and

$$s_i = s_{i-1} \text{ if } \phi_i = \phi_{i-1}$$

for all $i > 2$, with $s_1 = 1$.

For any cycle C in the digraph $D(A, s)$, let (v, u) be an edge in C such that (1) v has the smallest potential among all vertices in C, and (2) $\phi_u > \phi_v$. Such

---

[1] Afriat's orginal proof assumed that $a_{ij} \neq 0 \ \forall i \neq j$. This was relaxed by Diewert (1973) and Varian (1982).

[2] This proof is from Chung Piaw and Vohra (2003). The structure appears to have been overlooked (but used implicitly) in previous proofs – for example, Fostel, Scarf, and Todd (2007).

an edge exists, otherwise ϕ_i is identical for all vertices i in C. In this case, all edges in C have nonnegative edge weight in $D(A, s)$.

By choice, $\phi_u > \phi_v$. If $(v, u) \in S \cup Z$, then we have $\phi_u \leq \phi_v + w_{vu} \leq \phi_v$, which is a contradiction. Hence, $(v, u) \in T$. Now, note that all vertices q in C with the same potential as v must be incident to an edge (q, t) in C such that $\phi_t \geq \phi_q$. Hence the edge (q, t) must have nonnegative length, that is, $a_{qt} \geq 0$. Let p denote a vertex in C with the second smallest potential. Now, C has length

$$s_v a_{vu} + \sum_{(k,l) \in C, (k,l) \neq (v,u)} s_k a_{k,l} \geq s_v a_{v,u} + s_p(n-1) \min_{(i,j) \in S} \{a_{ij}\} \geq 0,$$

in other words, C has nonnegative length.

Because $D(A, s)$ is a digraph without any negative length cycles, $\ell(A)$ is feasible. ∎

> "There is no further information that can usefully or properly be added before bringing this chapter to an end."
>
> Harold Wilson

References

Afriat, S. N. "The Construction of a Utility Function from Expenditure Data," **International Economic Review**, 8, 67–77, 1967.

Ahuja, R. K., T. L. Magnanti and J. B. Orlin. "Network Flows: Theory, Algorithms and Applications," Prentice Hall, Upper Saddle River, NJ, 1993.

Anderson, E. J. and P. Nash. "Linear Programming in Infinite-Dimensional Spaces: Theory and Applications," John Wiley & Sons, Hoboken, NJ, 1987.

Archer, A. and R. Kleinberg. "Truthful Germs Are Contagious: A Local to Global Characterization of Truthfulness," Proceedings of the 9th ACM Conference on Electronic Commerce, 2008.

Arrow, K. J. "Social Choice and Individual Values," Wiley, New York, 1963.

Ashlagi, I., M. Braverman, A. Hassidim and D. Monderer. "Monotonicity and Implementability," manuscript, 2009.

Ausubel, L. and P. R. Milgrom. "Ascending Auctions with Package Bidding," **Frontiers of Theoretical Economics**, 1, 1–42, 2002.

Beaudry, P., C. Blackorby and D. Szalay. "Taxes and Employment Subsidies in Optimal Redistribution Programs," **American Economic Review**, 99(1), 216–242, 2009.

Berger, A., R. Müller and S. Naemi. "Characterizing Incentive Compatibility for Convex Valuations," **Lecture Notes in Computer Science**, 5814, 24–35, 2009.

Bikhchandani, S. "Ex Post Implementation in Environments with Private Goods," **Theoretical Economics**, 1, 369–393, 2006.

Bikhchandani, S., S. de Vries, J. Schummer and R. V. Vohra. "Linear Programming and Vickrey Auctions," in **Mathematics of the Internet**, edited volume of papers from an IMA topics workshop on Auctions and Markets. Co-edited by B. Dietrich and R. Vohra, Springer-Verlag, New York, 2002.

Bikhchandani, S., S. Chatterji, R. Lavi, A. Mu'alem, N. Nisan and A. Sen. "Weak Monotonicity Characterizes Incentive Deterministic Dominant Strategy Implementation," **Econometrica**, 74(4), 1109–1132, 2006.

Bikhchandani, S., S. de Vries, J. Schummer and R. V. Vohra. "Ascending Auctions for Integral (Poly)matroids with Concave Nondecreasing Separable Values," **Proceedings of the 19th ACM-SIAM Symposium on Discrete Algorithms**, 864–873, 2008.

Bikhchandani, S. and J. Ostroy. "The Package Assignment Model," **Journal of Economic Theory**, 377–406, 2002.

Blair, D. H. and R. A. Pollack. "Acyclic Collective Choice Rules," **Econometrca**, 50(4), 931–943, 1982.

Border, K. "Implementation of Reduced Form Auctions: A Geometric Approach," **Econometrica**, 59(4): 1175–1187, 1991.

Border, K. "Reduced Form Auctions Revisited," **Economic Theory**, 31(1), 167–181, 2007.

Branco, F. "Multiple Unit Auctions of an Indivisible Good," **Economic Theory**, 8(1), 77–101, 1996.

Chung, K. C. and J. Ely. "Ex-Post Incentive Compatible Mechanism Design," manuscript, 2001.

Chung, K. C. and W. Olszewski. "A Non-Differentiable Approach to Revenue Equivalence," **Theoretical Economics**, 2(4), 469–487, 2007.

Clarke, E. "Multipart Pricing of Public Goods," **Public Choice**, 19–33, 1971.

Cremer, J. and R. P. McLean. "Full Extraction of the Surplus in Bayesian and Dominant Strategy Auctions," **Econometrica**, 56(6), 247–257, 1988.

Dantzig, G. B. "Linear Programming and Extensions," Princeton University Press, Princeton, NJ, 1959.

Demange, G., D. Gale and M. Sotomayor. "Multi-Item Auctions," **Journal of Political Economy**, 94, 863–872, 1986.

de Vries, S., J. Schummer and R. V. Vohra. "On Ascending Vickrey Auctions for Heterogeneous Objects," **Journal of Economic Theory**, 132(1), 95–118, 2007.

Diewert, E. "Afriat and Revealed Preference Theory," **Review of Economic Studies**, 40, 419–426, 1973.

Duffin, R. J. "The Extremal Length of a Network," **Journal of Mathematical Analysis and Applications**, 5, 200–215, 1962.

Dutta, B., M. Jackson and M. Le Breton. "Strategic Candidacy and Voting Procedures," **Econometrica**, 69, 1013–1038, 2001.

Elkind, E. "Designing and Learning Optimal Finite Support Auctions," **Proceedings of the 18th ACM-SIAM Symposium on Discrete Algorithms**, 736–745, 2007.

Figueroa, N. and V. Skreta. "A Note on Optimal Allocation Mechanisms," **Economics Letters**, 102(3), 169–173, 2009.

Fostel, A., H. Scarf and M. J. Todd. "Two New Proofs of Afriat's Theorem," **Economic Theory**, 24(1), 211–219, 2007.

Gaertner, W. "Domain Conditions in Social Choice Theory," Cambridge University Press, New York, 2001.

Gershkov, A., B. Moldovanu, X. Shi. "Bayesian and Dominant Strategy Implementation Revisited," manuscript, 2011.

Green, J. and J-J. Laffont. "Characterization of Satisfactory Mechanisms for the Revelation of Preferences for Public Goods," **Econometrica**, 45, 727–738, 1977.

Groenvelt, H. "Two Algorithms for Maximizing a Separable Concave Function over a Polymatroid Feasible Region," **European Journal of Operational Research**, 54, 227–236, 1991.

Groves, T. "Incentives in Teams," **Econometrica**, 41, 617–631, 1973.

Gul, F. and E. Stachetti. "The English Auction with Differentiated Commodities," **Journal of Economic Theory**, 92, 66–95, 2000.

Gul, F. and E. Stachetti. "Walrasian Equilibrium with Gross Substitutes," **Journal of Economic Theory**, 87, 95–124, 1999.

Harris, M. and A. Raviv. "Allocation Mechanisms and the Design of Auctions," **Econometrica**, 49(6), 1477–1499, 1981.

References

Heydenreich, B, R. Müller, M. Uetz and R. V. Vohra. "A Characterization of Revenue Equivalence," **Econometrica**, 77(1), 307–316, 2009.

Holmström, B. "Groves' Scheme on Restricted Domains," **Econometrica**, 47(5), 1137–1114, 1979.

Hotelling, H. "Edgeworth's Taxation Paradox and the Nature of Demand and Supply Functions," **Journal of Political Economy**, 40, 577–616, 1932.

Iyenagar, G. and A. Kumar. "Optimal Procurement Mechanisms for Divisible Goods with Capacitated Suppliers," **Review of Economic Design**, 12, 129–154, 2008.

Jehiel, P., M. Meyer-ter-Vehn, B. Moldovanu and W. R. Zame. "The Limits of Ex Post Implementation," **Econometrica**, 74(3), 585–610, 2006.

Jehiel, P. and B. Moldovanu. "Efficient Mechanism Design with Interdependent Valuations," **Econometrica**, 69(5), 1237–1259, 2001.

Jehiel, P., B. Moldovanu and E. Stachetti. "Multidimensional Mechanism Design for Auctions with Externalities," **Journal of Economic Theory**, 85, 258–293, 1999.

Kagel, J. H. and D. Levin. "Behavior in Multi-Unit Demand Auctions: Experiments with Uniform Price and Dynamic Vickrey Auctions," **Econometrica**, 69, 413–454, 2001.

Kalai, E. and E. Muller. "Characterization of Domains Admitting Nondictatorial Social Welfare Functions and Nonmanipulable Voting Procedures," **Journal of Economic Theory**, 16(2), 457–469, 1977.

Kelso, A. S. and V. P. Crawford. "Job Matching, Coalition Formation, and Gross Substitutes," **Econometrica**, 50, 1483–1504, 1982.

Krishna, V. and E. Maenner. "Convex Potentials with an Application to Mechanism Design," **Econometrica**, 69(4), 1113–1119, 2001.

Krishna, V. and M. Perry. "Efficient Mechanism Design," manuscript, 2000.

Laffont, J. J. and J. Robert. "Optimal Auction with Financially Constrained Buyers," **Economic Letters**, 52(2), 181–196, 1996.

Lavi, R., A. Mu'alem and N. Nisan. "Towards a Characterization of Truthful Combinatorial Auctions," **Proceedings of the 44th Annual IEEE Symposium on Foundations of Computer Science**, 2003.

Lavi, R., A. Mu'alem and N. Nisan. "Two Simplified Proofs for Roberts' Theorem," manuscript, 2004.

Lovejoy, W. "Optimal Mechanisms with Finite Agent Types," **Management Science**, 52(5), 788–803, 2006.

Malakhov, A. and R. V. Vohra. "An Optimal Auction for Capacity Constrained Bidders: A Network Perspective," **Economic Theory**, 39(1), 113–128, 2009.

Malakhov, A. and R. V. Vohra. "Single and Multi-Dimensional Optimal Auctions – a Network Approach," manuscript, September 2004.

Manelli, A. and D. Vincent. "Bayesian and Dominant Strategy Implementation in the Independent Private Values Model," **Econometrica**, 78(6), 1905–1938, 2010.

Manelli, A. and D. Vincent. "Multidimensional Mechanism Design: Revenue Maximization and the Multiple-Good Monopoly," **Journal of Economic Theory**, 137, 153–185, 2007.

Mas-Collel, A., M. Whinston and J. Green, "Microeconomic Theory," Oxford University Press, New York, 1995.

McAfee, R. P. and J. McMillan. "Multidimensional Incentive Compatibility and Mechanism Design," **Journal of Economic Theory**, 46, 335–354, 1988.

Meyer-ter-Vehn, M. and B. Moldovanu. "Ex-Post Implementation with Interdependent Valuations," manuscript, 2002.

Mezzetti, C. 2004. "Mechanism Design with Interdependent Valuations: Efficiency," **Econometrica**, 72(5), 1617–1626, 2004.

Mierendorf, K. "Asymmetric Reduced Form Auctions," manuscript, 2009.

Milgrom, P. "Assignment Messages and Exchanges," **American Economic Journal: Microeconomics**, 1(2), 95–113, 2009.

Milgrom, P. and I. Segal. "Envelope Theorems for Arbitrary Choice Sets," **Econometrica**, 70(2), 583–601, 2002.

Morris, S. "The Common Prior Assumption in Economic Theory," **Economics and Philosophy**, 11, 227–253, 1995.

Müller, R., A. Perea and S. Wolf. "Weak Monotonicity and Bayes-Nash Incentive Compatibility," **Games and Economic Behavior**, 61, 344–358, 2007.

Murota, K. "Discrete Convex Analysis," **SIAM Monographs on Discrete Mathematics and Applications, Society for Industrial and Applied Mathematics (SIAM)**, Philadelphia, PA, 2003.

Murota, K. "Submodular Function Minimization and Maximization in Discrete Convex Analysis," manuscript, 2008.

Myerson, R. B. "Optimal Auction Design," **Mathematics of Operations Research**, 6(1), 58–73, 1981.

Myerson, R. B. Revelation Principle, "The New Palgrave Dictionary of Economics," Second Edition. Eds. S. N. Durlauf and L. E. Blume. Palgrave Macmillan, New York, 2008.

Myerson, R. B. and M. Satterthwaite. "Efficient Mechanisms for Bilateral Trading," **Journal of Economic Theory**, 28, 265–281, 1983.

Nemhauser, G. L. and L. A. Wolsey. "Integer and Combinatorial Optimization," John Wiley & Sons, New York, 1988.

Nisan, N and I. Segal. "The Communication Requirements of Efficient Allocations and Supporting Prices," **Journal of Economic Theory**, 129, 192–224, 2006.

Pai, M. and R. V. Vohra. "Optimal Auctions with Financially Constrained Bidders," manuscript, 2008.

Parkes, D. and L. Ungar. " i-Bundle: An Efficient Ascending Price Bundle Auction," **Proceedings of the 1st ACM Conference on Electronic Commerce**, 148–157, 1999.

Piaw, T. C. and R. V. Vohra. "Afriat's Theorem and Negative Cycles," manuscript, 2003.

Rahman, D. "Detecting Profitable Deviations," manuscript, 2009.

Reny, P. J. "Arrow's Theorem and the Gibbard-Satterthwaite Theorem: A Unified Approach," **Economics Letters**, 70(1), 99–105, 2001.

Riley, J. G. and W. F. Samuelson. "Optimal Auctions," **The American Economic Review**, 71(3), 381–392, 1981.

Roberts, K. "The Characterization of Implementable Choice Rules," in Jean-Jacques Laffont, editor, "Aggregation and Revelation of Preferences." **Papers presented at the 1st European Summer Workshop of the Econometric Society**, 321–349, North-Holland, 1979.

Rochet, J. C. "A Necessary and Sufficient Condition for Rationalizability in a Quasilinear Context," **Journal of Mathematical Economics**, 16, 191–200, 1987.

Rochet, J. C. and P. Choné. "Ironing, Sweeping, and Multidimensional Screening," **Econometrica**, 66, 783–826, 1998.

Rochet, J. C. and L. A. Stole. "The Economics of Multidimensional Screening," in *Advances in Economics and Econometrics: Theory and Applications, Eighth World Congress*, ed. M. Dewatripont, L. P. Hansen and S. J. Turnovsky. Cambridge University Press, Cambridge, 2003.

References

Rockafellar, R. T. "Characterization of the Subdifferentials of Convex Functions," **Pacific Journal of Mathematics**, 17(3), 487–510, 1966.

Rockafellar, R. T. "Convex Analysis," Princeton University Press, Princeton, NY, 1970.

Rockafellar, R. T. and R. J-B. Wets. "Variational Analysis," Springer-Verlag, New York, 1998.

Saari, D. "Capturing the Will of the People," **Ethics**, 113(2), 333–349, 2001.

Saari, D. "Chaotic Elections! A Mathematician Looks at Voting," American Mathematical Society, Providence, RI, 2001.

Saks, M. and L.Yu. "Weak Monotonicity Suffices for Truthfulness on Convex Domains," **Proceedings of the 6th ACM Conference on Electronic Commerce** (EC05), 286–293, 2005.

Sato, K. "A Study on Linear Inequality Representation of Social Welfare Functions," masters thesis, University of Tsukuba, 2006.

Serrano, R. "The Theory of Implementation of Social Choice Rules," **SIAM Review**, 46(3), 377–414, 2004.

Sethuraman, J., C. P. Teo and R. Vohra. "Anonymous monotonic Social Welfare Functions," **Journal of Economic Theory**, 128(1), 232–254, 2006.

Sethuraman, J., C. P. Teo and R. Vohra. "Integer Programming and Arrovian Social Welfare Functions," **Mathematics of Operations Research**, 28(2), 2003.

Shapley, L. S. and M. Shubik. "The Assignment Game I: The Core," **International Journal of Game Theory**, 1, 111–130, 1972.

Varian, H. "The Non-parametric Approach to Demand Analysis," **Econometrica**, 50, 945–974, 1982.

Vickrey, W. "Counterspeculation, Auctions, and Competitive Sealed Tenders," **Journal of Finance**, 16, 8–37, 1961.

Vohra, R. V. "Advanced Mathematical Economics," Routledge, Oxford 2005.

Walras, L. "Elements of Pure Economics, or the Theory of Social Wealth," (translated by W. Jaffe and Richard D. Irwin, 1954), 1874.

Wilson, R. "Nonlinear Pricing," Oxford University Press, Oxford 1993.

Index

2-cycle condition, 45
2-cycle inequality, 44

active bidders, 91
affine maximizers, 54
Afriat's Theorem, 162
agents are substitutes condition, 88
allocation network, 42
allocation rule, 38
arcs, 25
Arrovian social welfare function, 8
Arrow's Impossibility Theorem, 13
ASC, 88
Ascending Auctions, 89

Bayes-Nash incentive compatible mechanism (BNIC), 77
Bayesian incentive compatibility, 77
Border's Theorem, 121

Candidate Stability, 18
Combinatorial Auctions, 82
complementary slackness, 90
complete, 24
Condorcet, 9
cone assumption, 130
connected, 25
Core, 87
Correlated types, 130
cross at most once, 76
cycle, 25
cycle elimination, 10
cyclic monotonicity, 72

decisive, 8
decisiveness implications, 10
decomposed, 30
decomposition monotone, 53
degree, 24
demand correspondence, 91
demand rationing, 65
diagonal BNIC constraints, 151
dictatorial, 8
direct mechanism, 22
directed, 25
dominant strategy equilibrium, 21
dominant strategy mechanism, 40
dot-product valuations, 39
downward BNIC constraint, 115

edges, 24
efficient allocation, 79
English clock auction, 81
ex-post incentive compatible, 74
expected allocation, 114
extended formulations, 84

feasible flow, 29
flow decomposition, 31
full information problem, 113

generalized axiom of revealed preference, 162
Gibbard-Satterthwaite Theorem, 17
graph, 24
Gross Substitutes, 105
gross substitutes, 107

Index

implementable in Bayes-Nash equilibrium (IBN), 77
combine into one entry, 40
implementable in dominant strategies (IDS), 40
Independence of Irrelevant Alternatives, 8
individual rationality (IR), 131
infinite dimensional linear program, 140
interdependent, 73
interim allocation, 114
ironed virtual value, 127

knapsack problem, 157

locally nonsatiated, 161
loop, 24

Majority rule, 9
marginal product, 81
mechanism, 21
minimally undersupplied, 94
monotone hazard condition, 118
Monotonicity, 15
monotonicity, 45
Muller-Satterthwiate Theorem, 16

network, 26
Newtonian, 69
node-arc incidence matrix, 28
nodes, 24

overdemand, 93

Pareto Optimality, 15
path, 24
payment rule, 38
Polyhedral approach, 121
polymatroid, 125
polymatroid optimization, 125
pooling, 126
preference domain, 7
price adjustment, 95
price path, 107

Primal-Dual Algorithm, 89
profiles, 7
profitable, 42

quasiefficient, 127

rationalizability, 160
regular, 60
restricted primal, 90
Revelation Principle, 22
revenue equivalence, 59
Roberts' Theorem, 54

shortest path polyhedron, 33
signal, 74
simple cycle, 25
simple path, 24
social choice function, 14
social welfare function, 7
spanning tree, 25
splittable, 62
Strategy-Proof, 15
strongly connected, 26
subgradient algorithms, 99
submodular, 125
supply correspondence, 91

taxation principle, 42
top single crossing, 76
Transitivity, 9
truthfully implementable, 22
two-cycle connected, 63
type, 20
type space, 20

Unanimity, 8, 18
undersupplied, 94
undetectable, 42
uniformly ordered, 76, 134
upward BNIC constraint, 115

Vickrey-Clarke-Groves (VCG), 81
virtual valuation, 118

Walrasian tâtonnement, 82

Other titles in the series (continued from page iii)

Eric Ghysels, Norman R. Swanson, and Mark Watson, Editors, *Essays in econometrics: Collected papers of Clive W. J. Granger* (Volume I), 978 0 521 77297 6, 978 0 521 77496 3

Eric Ghysels, Norman R. Swanson, and Mark Watson, Editors, *Essays in econometrics: Collected papers of Clive W. J. Granger* (Volume II), 978 0 521 79207 3, 978 0 521 79649 1

Cheng Hsiao, *Analysis of panel data, second edition*, 978 0 521 81855 1, 978 0 521 52271 7

Mathias Dewatripont, Lars Peter Hansen, and Stephen J. Turnovsky, Editors, *Advances in economics and econometrics – Eighth World Congress* (Volume I), 978 0 521 81872 8, 978 0 521 52411 7

Mathias Dewatripont, Lars Peter Hansen, and Stephen J. Turnovsky, Editors, *Advances in economics and econometrics – Eighth World Congress* (Volume II), 978 0 521 81873 5, 978 0 521 52412 4

Mathias Dewatripont, Lars Peter Hansen, and Stephen J. Turnovsky, Editors, *Advances in economics and econometrics – Eighth World Congress* (Volume III), 978 0 521 81874 2, 978 0 521 52413 1

Roger Koenker, *Quantile Regression*, 978 0 521 84573 1, 978 0 521 60827 5

Charles Blackorby, Walter Bossert, and David Donaldson, *Population issues in social choice theory, welfare economics, and ethics*, 978 0 521 82551 1, 978 0 521 53258 7

John E. Roemer, *Democracy, Education, and Equality*, 978 0 521 84665 3, 978 0 521 60913 5

Richard Blundell, Whitney K. Newey, and Thorsten Persson, *Advances in economics and econometrics – Ninth World Congress* (Volume I), 978 0 521 87152 5, 978 0 521 69208 3

Richard Blundell, Whitney K. Newey, and Thorsten Persson, *Advances in economics and econometrics – Ninth World Congress* (Volume I), 978 0 521 87153 2, 978 0 521 69209 0

Richard Blundell, Whitney K. Newey, and Thorsten Persson, *Advances in economics and econometrics – Ninth World Congress* (Volume I), 978 0 521 87154 9, 978 0 521 69210 6

Fernando Vega-Redondo, *Complex social networks*, 978 0 521 85740 6, 978 0 521 67409 6

Itzhak Gilboa, *Theory of decision under uncertainty*, 978 0 521 51732 4, 978 0 521 741231

Krislert Samphantharak and Robert M. Townsend, *Households as corporate firms: an analysis of household finance using integrated household surveys and corporate financial accounting*, 978 0 521 19582 9, 978 0 521 12416 4

CPSIA information can be obtained at www.ICGtesting.com
Printed in the USA
LVOW090020310512

284023LV00002B/1/P